ASIA'S SPACE RACE

Contemporary Asia in the World

CONTEMPORARY ASIA IN THE WORLD

David C. Kang and Victor D. Cha, Editors

This series aims to address a gap in the public-policy and scholarly discussion of Asia. It seeks to promote books and studies that are on the cutting edge of their respective disciplines or in the promotion of multidisciplinary or interdisciplinary research but that are also accessible to a wider readership. The editors seek to showcase the best scholarly and public-policy arguments on Asia from any field, including politics, history, economics, and cultural studies.

Beyond the Final Score: The Politics of Sport in Asia, Victor D. Cha, 2008

The Power of the Internet in China: Citizen Activism Online, Guobin Yang, 2009

China and India: Prospects for Peace, Jonathan Holslag, 2010

India, Pakistan, and the Bomb: Debating Nuclear Stability in South Asia, Šumit Ganguly and S. Paul Kapur, 2010

Living with the Dragon: How the American Public Views the Rise of China, Benjamin I. Page and Tao Xie, 2010

East Asia Before the West: Five Centuries of Trade and Tribute, David C. Kang, 2010

Harmony and War: Confucian Culture and Chinese Power Politics, Yuan-Kang Wang, 2011

Strong Society, Smart State: The Rise of Public Opinion in China's Japan Policy, James Reilly, 2012

ASIA'S SPACE RACE

National Motivations, Regional Rivalries, and International Risks

JAMES CLAY MOLTZ

COLUMBIA UNIVERSITY PRESS   *NEW YORK*

Columbia University Press
Publishers Since 1893
New York   Chichester, West Sussex
cup.columbia.edu
Copyright © 2012 Columbia University Press
Paperback edition, 2018
Library of Congress Cataloging-in-Publication Data
Moltz, James Clay.
Asia's space race : national motivations, regional rivalries, and international risks /
James Clay Moltz.
p.   cm. — (Contemporary Asia in the world)
Includes bibliographical references and index.
ISBN 978-0-231-15688-2 (cloth) — ISBN 978-0-231-15689-9 (pbk.)
ISBN 978-0-231-52757-6 (e-book)
1. Space race—Asia.   2. Astronautics and state—Asia.   3. Astronautics and state—United
States.   I. Title.   II. Series.
TL789.8.A78M65   2011
629.4'1095—dc22   2011004616

Book and cover design by Milenda Nan Ok Lee
Cover image: Gong hui/Imaginechina

CONTENTS

PREFACE

The subject of space competition in Asia remains highly topical. Since this book's original publication in 2012, much has happened. Japan has increased its cooperation with the United States in such areas as space situational awareness, while increasing its military space budget and developing new civilian technologies, including the low-cost Epsilon launcher. China has continued to expand and modernize its space infrastructure: opening its Hainan Island launch site; putting into operation the Beidou position, timing, and navigation system; and expanding its constellations of civil and military satellites. China has also conducted additional tests of anti-satellite weapons. India has launched a successful Mars probe, expanded its military space cadre, and won contracts for U.S. commercial satellite launches. South Korea successfully launched its first satellite in 2013 on a joint-venture rocket with Russia, while also advancing its own commercial satellite-manufacturing business and working toward a larger, domestically produced, space-launch vehicle. Meanwhile, North Korea has launched two satellites—in 2012 and 2016—but seems to have recently focused on its military missiles rather than space development, thus raising security tensions in the region. Other countries, including Australia, have developed more ambitious national space policies. New Zealand in particular has put itself on the map by opening a space launch facility, which hosts Rocket Lab's Electron launcher. All of these developments provide evidence of the continued expansion of Asian space activity.

In issuing this paperback edition, I have not changed the arguments or updated the various chapters. The history and domestics contexts of these space programs and the nature of their existing rivalries have not changed, although the countries have continued to develop new space technologies and programs. My hope is that this book remains valuable as a foundational study for understanding subsequent Asian space developments.

James Clay Moltz
December 2017

ACKNOWLEDGMENTS

The idea for this book emerged out of a project that John Logsdon of George Washington University's Space Policy Institute and I organized in 2006 and 2007—with the support of the John D. and Catherine T. MacArthur Foundation—to study regional perspectives of space security in Europe and Asia. It became clear from our workshop in Paris in May 2006, held in cooperation with Xavier Pasco and the Foundation pour la Researche Strategique, that European countries tended to share a consensus view on space security questions geared toward "human security" and implemented through a set of collaborative and often collective policies toward space. By contrast, our meeting held in Tokyo in April 2007, in cooperation with Masashi Nishihara and the Research Institute for Peace and Security, highlighted the largely inchoate nature of Asian space relations, the relative absence of cooperation, and the lack of consensus among major powers on space security. Instead, weak regional institutions, mutual mistrust, and poor and sporadic communications about space characterized their relations.

The problem posed by this situation is that the rapid *rise* of Asian space activity presages a competitive and possibly conflict-prone environment. Developments such as China's antisatellite test in January 2007, the *Chang'e* research program, and its human spaceflight missions; Japan's successful conduct of its *Kaguya* lunar project and its evolving activities associated with the *Kibo* module on the *International Space Station* (*ISS*); India's *Chandaryaan-1* and *Chandaryaan-2* lunar probes; and South Korea's satellite progress, astronaut mission to the *ISS*, and plans to establish an independent space-launch capability all provide evidence that an increasing portion of early twenty-first-century space activities will be conducted by Asian countries. Yet few studies have tried to explain these dynamics from a regional perspective or by noting the increasing trends toward competition and possible conflict. This book, begun in the spring of 2008, has been an effort to examine Asian space dynamics from a domestic, regional, and international context across the full range of space activities: civil, commercial, scientific, and military. Its goal has been to determine where these countries are headed and what the implications might be, both for the United States and for the broader space community. In the end, my conclusion is that regional space conflict is avoidable but will require considerable preventative action not yet in evidence in Asia.

This book would not have been possible without support from the Naval Postgraduate School's Office of Sponsored Research, which provided me with two academic quarters of research and travel funding through the Research Initiation Program. This included trips to a number of international conferences to meet and discuss emerging issues with experts from the region and travel to South Korea, Japan, and China for more focused background interviews. In each of these countries, I incurred personal debts of gratitude to various individuals who assisted my research and greatly improved the product that became this book.

In South Korea, I benefited from assistance provided by my friend and former colleague Dr. Daniel Pinkston, the director of the International Crisis Group's office in Seoul. He provided useful ideas, and his office manager, Ja-Young Won, was kind enough to help facilitate a number of meetings. In terms of interviews, I am especially grateful to Director Chin-Young Hwang and his colleagues Jong-Taek Huh and Seorim Lee in the Policy and International Relations Division at the Korean Aerospace Research Institute (KARI) for hosting me in Taejon, providing a fascinating tour, and answering all of my questions. I also thank Kyungmin Kim of Hanyang University, Yong-sup Han and Hong-je Cho of the Korean National Defense University, Taewoo Kim of the Korea Institute for Defense Analyses, and Jangkeun Lee of the Ministry of Foreign Affairs and Trade for agreeing to meet with me to share their insights on South Korea's space policy and their perspectives on the role of space more generally in Northeast Asia. I also thank the students at Hanyang University and the officers at the Korean National Defense University for their questions on presentations I gave there. I was later fortunate enough to meet with KARI's vice president, Dr. Jeong-Joo Rhiu, and with the South Korean astronaut So-yeon Yi at the NASA-Ames Research Center in Mountain View, California.

In Japan, I benefited greatly from the generous advice of Professor Setsuko Aoki of Keio University, who helped me pinpoint leading officials and analysts from her extensive set of contacts with whom I should meet—and who even undertook several personal introductions on my behalf. I also thank NASA's Gib Kirkham for his advice on other officials and experts and for our general discussions about space activity in the region. I am grateful to the following individuals for agreeing to meet with me and share their views on Japan's current and future space trajectory: Takayuki Kawai, Takashi Endo, Masahiro Saitoh, and Jun'ichiro Kawaguchi, all of the Japan Aerospace Exploration Agency; Fumihiko Toshida of *Asahi Shimbun*; Masakazu Toyoda and Norihiko Saeki of the Cabinet Secretariat's Headquarters for Space Policy; Dan Cintron from the Political-Military Affairs Section of the U.S. Embassy in Japan; Diet Member Katsuyuki Kawai; Katsuhisa Furukawa of the Research Institute of Science and Technology for Society; Masashi Nishihara of the Research Institute for Peace and Security; Kazuto Suzuki of Hokkaido University; Yoichi Iida of the Ministry of Economy, Trade, and Industry; Lance Gatling of Gatling Associates; Jun Yanagi, Kei Narisawa, and Shigeru Umetsu of the Japanese Ministry of Foreign Affairs; and Setsuko Aoki (an expert in her own right and member of the Eminent Persons' Group, which was set up in 2008 to advise the Cabinet Secretariat on space activities). In addition, I thank those in the audience who raised questions during a presentation I gave at Keio University, especially those from Japanese industry and the media. Professor Suzuki also very kindly agreed to read my final chapter on Japan and provided useful comments and corrections.

My work on China benefited from a number of contacts. First, I am very grateful to Professor Li Bin of Tsinghua University for providing me with an invitation to China, answering the many questions I had, and providing contacts among Chinese academics. At the Beijing University of Aeronautics and Astronautics, I thank Professors Huang Hai and Chen Shenyan and about twenty of their students who met with me. I am also grateful to my former Monterey Institute of International Studies colleague Yuan Jing-dong for helping contact space policy experts at the Chinese Foreign Ministry, where I met with Li Song and Zhang Ze, both of the Arms Control and Disarmament Department. Our meeting in Beijing was a nice re-union, as I had advised Li Song on a research paper nearly ten years before in Monterey but had not seen him since, during which time he had risen within the ministry. At the Institute of Security and Strategic Studies under the China Institutes of Contemporary International Relations, I met with Professor Guo Xiaobing and his colleague Liu Chong. I am grateful to Geng Kun of the Great Wall Industry Corporation for arranging a very useful discussion with Vice President Gao Ruofei and several other col-leagues, including Zhang Cunling, Li Honggui, and Liu Bo. I much appre-ciated their willingness to spend several hours answering my questions. In all, the Chinese experts and officials I met with were unfailingly helpful and courteous. In the United States, I am grateful for the counsel of Pat-rick Besha at NASA, whose expert knowledge of China's space program provided some useful pointers before and after my visit to Beijing. I also thank Eric Hagt for his advice and suggestions. Kevin Pollpeter's highly informed comments on my final chapter on China improved it signifi-cantly and prevented some embarrassing mistakes. My Naval Postgradu-ate School (NPS) colleague Chris Twomey merits recognition for his gener-osity in reading several draft chapters in their early form and for providing additional comments on the final Chinese chapter.

In regard to India, limits of funding and time prevented me from making a visit to conduct on-the-ground meetings. Fortunately, Deviprasad Karnik of the Indian Space Research Organization (ISRO) provided me with an extremely useful introduction to India's current space activities in a meet-ing held in his office at the Indian Embassy in Washington in November 2008. I also benefited from meetings in California in the spring of 2009 with Bharath Gopalaswamy, then a postdoctoral fellow at Cornell Univer-sity, and Amandeep Gill of the India Foreign Ministry, then a visiting fel-low at Stanford University's Center for Security and Cooperation. I am very grateful as well for a discussion with NASA's Garvey MacIntosh, who helped round out my understanding of ISRO's cooperative efforts. Fortu-nately, the relative availability of detailed materials in English on the In-dian space program assisted me greatly in completing this chapter, as did contacts with experts such as Ajey Lele and K. K. Nair.

In addition, I thank my former colleagues—particularly Dr. William Marshall and Robert Schingler—in the Space Futures Working Group at the NASA Ames Research Center, where we discussed a number of related themes and where I benefited from initial feedback on the project as it began. In the U.S. government, I am grateful to Richard Buenneke of the State Department, for his informed insights and critiques, and to Dr. Peter Hays, formerly of the Defense Department's National Security Space Office, for his willingness to listen and provide honest comments. David Arnold and Matt Tritle deserve thanks for helping get skeptical Pentagon administrators to sign off on my research trip to China.

Beyond those mentioned above who commented on the manuscript, Dr. James Vedda of the Aerospace Corporation generously provided both a detailed review of the complete manuscript and his formidable intellectual backing for the project. In addition, two anonymous reviewers from Columbia University Press added useful pointers and suggestions. My composite chapter that attempts to review ten of Asia's developing space programs would not have been nearly as accurate without a careful critique from the Australian space expert Brett Biddington. I greatly appreciated his timely comments.

For moral support, I thank Harold Trinkunas, my department's former chairman, who instantly liked the topic of the book when I mentioned it to him and helped convince me to pursue this research. Professors Tom Bruneau, Dan Moran, and Doug Porch all provided much-appreciated encouragement. I owe a weighty dose of gratitude to my wife, Sarah Diehl, who endorsed the project and tolerated my long absences from home during the research stage and many lost weekends as I worked to complete the manuscript. She will, perhaps, be even happier than me to see this book in print. I also thank my colleagues in the NPS Space Systems Academic Group, especially Rudy Panholzer, for their support and interest in my work. The Knox Library at NPS and its staff proved to be another great benefit, as I was able to find a number of sources on Asian space developments previously untapped by past Western scholars. My students helped stimulate my thinking and interest in this topic as well.

In terms of Asian names, their transliteration, and their word order, I have opted to follow a flexible policy tied to conventional use or the style used by an author in printed publications. This has resulted in some unavoidable inconsistencies (Syngman Rhee versus Park Chung-hee), but should be relatively easy to follow.

In the end, regrettably, I know mistakes of fact and judgment will be discovered in these pages. I hope they are small and posit them to my own imperfect abilities rather than to any fault by the many people who helped me. My only excuse is that this book covers a large and complicated subject that is also a moving target. The ideas expressed here are my personal interpretations and opinions and do not represent the official policies of the U.S. Navy or the U.S. Department of Defense. Still, I hope I have helped outline a set of new issues worthy of greater attention by students, the interested public, space experts, and government officials in Asia and around the world. More work will certainly be needed in the policy realm in the coming years to ensure that Asia's dynamic foray into the area of space activity remains peaceful and, ultimately, serves to promote greater international cooperation in developing the space environment.

ASIA'S SPACE RACE

ASIA'S NEW PRESENCE IN SPACE

Asia's international relations have moved increasingly to the forefront of global attention in recent years. The region's greater openness and expanding economic might have led to greater Asian integration into the international marketplace and increasing Asian influence relative to other world actors. These trends have been caused and stimulated by Asian efforts to be recognized globally in science and technology. Space activity is part of these processes.

In some Asian countries, unfortunately, these trends have spurred regional competition, nationalism, and arms acquisition. As a result, Asia is witnessing a growing rivalry for technological ascendancy, political prestige, and security advantages through new space capabilities. Current trends exhibit a variety of action-reaction dynamics that may eventually become difficult to control, if efforts are not made to adopt confidence-building measures, increase transparency, and initiate consultations to prevent future space conflict. In discussing the emerging competition among China, India, and Japan for regional space leadership, the Asia watcher Bill Emmot argues "all three believe that space will be the next military battleground, and all three believe that to be both safe and powerful they need to develop their own space technology."[1]

Asian trends stand in sharp contrast to space developments in Europe, where the leading nations cooperate extensively. Eighteen countries in that region are now engaged in close practical integration within the jointly financed European Space Agency (ESA).[2] With a yearly budget of over $5 billion, ESA carries out coordinated research and development and implements a range of significant space projects for scientific and commercial purposes. Moreover, although these countries frequently wrestle over ESA budgetary matters,[3] they share consensus views on space security, the need for collective approaches to problem solving, and reliance on legal remedies instead of warfare.[4] For example, the European countries have released a model "code of conduct" for space in hopes of preventing future conflict and increasing global space cooperation. They are now moving forward to attract broader international support.

By contrast, Asia's space powers are largely isolated from one another, do not share information, and display a tremendous *divergence* of perspectives regarding their space goals and a tendency to focus on *national* solutions to space challenges and policies of self-reliance rather than on regionwide policies or multilateral approaches.[5] As the Japanese analyst Setsuko Aoki observes, "the foundation for Asian collective security in space is fragile, if not non-existent."[6] Such hostile dyads as India-China, China-Japan, India-Pakistan, Japan-South Korea, and North Korea-South Korea indicate that Asian countries see space largely as an extension of other competitive realms and are carefully watching regional rivals, attempting to match or at least to check their capabilities, influence, and power.

China's dramatic entrance into civil, commercial, and military space activities over the past decade has played a major role in putting Asia's other space players on notice. A number of rivals have expanded their space programs as a result, in some cases dramatically. India has enhanced its satellite reconnaissance capabilities, established a military space command, and launched its first in a series of planned lunar probes, while hinting at future human spaceflights to match China's. Japan has invested in costly lunar missions, maintains an active astronaut corps, and has built new modules and transport spacecraft for the *International Space Station* (*ISS*). Of special concern to its neighbors is Japan's decision to scrap its forty-year-old Diet Law on space activity and to allow military uses for the first time. South Korea, meanwhile, has moved forward with advanced imaging, communications, and scientific satellites, has purchased space-launch technology from Russia, and paid for a South Korean astronaut's flight to the *ISS*. Pakistan, North Korea, and Indonesia are all aiming for independent space-launch capability as well, and Australia, Malaysia,

Singapore, Taiwan, Thailand, and Vietnam are all working to expand their space capabilities. Although Asian leaders do not like to speak in these terms, all of the evidence points to the emergence of a "space race" in the region. But, unlike the cold war's one-hundred-yard dash to the Moon, Asia's space rivalry is more like a long-duration cross-country race, with many runners and many different objectives. Some know they will not win, but they want others (and their own populations) to know that they are actively participating. Fortunately, Asia's space competition does not yet emphasize weapons systems. Nevertheless, China has tested such capabilities, India has announced plans to do so, and Japan and South Korea have missile-defense systems that could in the future be adapted to hit space objects. In this context, the relative absence of cooperation and trust among Asia's main space powers raises the possibility of future instability if new management mechanisms are not developed and put into place before further space weapons capabilities are fully developed, tested, and deployed.

Some countervailing trends toward Asian space cooperation do exist. Asian space powers have all joined the 1967 Outer Space Treaty and support United Nations resolutions promoting space transparency and PAROS, the Prevention of an Arms Race in Outer Space. The bulk of current space spending in Asia goes to civilian applications and space science as well as, in a few countries, human spaceflight. Asian countries participate in a range of international cooperative projects with non-Asian countries and have established two regional space cooperation organizations (one led by China and one led by Japan), which have limited—but growing—training, data exchange, and operational activities. Finally, in the past few years, some cooperation has even occurred between erstwhile rivals. Japan provided remote-sensing data to China after the deadly Sichuan earthquake in 2008, and there are several efforts underway to increase regional cooperation in tsunami prediction, disaster relief, and climate monitoring operations. Still, these activities remain the exception, and military uses of space are increasing. Increasing nationalism is also evident in Asia's space programs, as countries vie to achieve civil and scientific firsts ahead of their neighbors, using space accomplishments to score points within deep-seated historical rivalries.

The United States and the rest of the international community have a major stake in the future stability and security of space. Since space is an interactive environment, what Asian countries do will affect the ability of all countries to carry out peaceful activities in space, including military support functions. Since China's antisatellite test in January 2007, debates about the future of space security now focus increasingly on risks posed by space debris, possible new Chinese military activities, and the possibility of countermoves in space by India or others. On November 2, 2009, an article in the *People's Liberation Army Daily* by the head of China's air force, General Xu Qiliang, wrote of the "historical inevitability" of military conflict becoming extended into space. Although similar views have been expressed by a number of senior U.S., Russian, and Indian military officers, Xu's remarks created an uproar in Washington. The China analyst Gordon Chang argued in the pages of *Forbes* magazine that the Chinese military has now "announced its intention to begin the space arms race in earnest."[7] Chang spoke of China's goal "to dominate space" and predicted a "brutal competition" with the United States.[8]

Unlike in January 2007, however, when China remained silent for two weeks after conducting a destructive antisatellite test, the Chinese government reacted quickly to the general's 2009 comments. The Chinese Foreign Ministry issued an official statement distancing the government from General Xu's remarks and reaffirming China's peaceful intentions in space.[9] Such statements seem to lend credence to predictions from the "engagement" school of American China watchers, who have suggested that Beijing's growing integration into the international economy will eventually limit its adventurism, because it now has much more to lose in a conflict with the United States (or other nations) than it did in prior decades, when it was less a part of the world economy.[10] Still, many U.S. observers and officials alike worry about existing trends and feel that the passage of time may simply lead to growing Chinese capabilities without restraints, to the detriment of the United States and other nations, rather than creating new forms of interdependence.

Near the end of the George W. Bush administration, then-NASA Administrator Michael D. Griffin sent a memorandum to the heads of the U.S. National Security Council, the Office of Management and Budget, and the White House's Office of Science and Technology Policy outlining the serious challenge he perceived from China in space.[11] Griffin's memo leaked out to the Internet and caused a minor scandal, given its robust assessment of China's space capabilities and the NASA chief's palpable fear that the United States was falling behind in an emerging twenty-first-century space competition. Griffin lamented the limits of NASA funds and predicted, "The next decade will be a period in which U.S. primacy in space will already be in question."[12] He went on: "it is my considered judgment that China will have the technical and budgetary wherewithal to conduct a manned mission to the surface of the moon before the United States plans to return."[13] Griffin warned of significant international implications from such an event, saying it would have an "enormous, and not fully predictable, effect on global perceptions of U.S. leadership in the world."[14]

Given its parallel civil, commercial, and military activities in space, the United States is conflicted about how to deal with China's emergence—whether through positive engagement, military measures, or balancing with allies (such as Japan and South Korea) or friends (such as India). Fortunately, the military dimensions of Asia's emerging space competition are not yet set in stone. It remains possible that U.S.-Chinese military space confrontation can be avoided. But this will require a deviation from current trajectories and a conscious effort by all states concerned.

ASIA AS SPACE LEADER?

Historians generally believe that Asian countries—particularly China—led the world in discovering gunpowder and in undertaking a series of experiments from at least the 1200s toward developing self-propelled arrows, related incendiaries, and early rockets.[15] While details in the historical record remain limited, it is clear that a number of Asian civilizations—including the Chinese, Khmer, Thai, Lao, Korean, and Indian—used various forms of small rockets and incendiaries from the thirteenth to nineteenth centuries both in celebrations and in battle. But Western advances on these devices—including British Congreve rockets developed in the late 1700s and early 1800s—eventually surpassed Asian weapons. In the twentieth century, the West surged ahead through development of liquid-fuel engines, cryogenic propellants, and strong but lightweight rocket bodies that provided far greater reliability, distance, and lethality. In the late twentieth century, Asian countries—including China, India, and Japan—began to borrow back from the West in developing reliable modern rockets for space use and, eventually, independent access to low-Earth, geostationary, and lunar orbits.

Most Asian countries are still net recipients of space technologies and few have led international projects to date. More generally, because of on-going military disputes among the powers and recent problems with the largely state-led development models used by most of them, Asia's current space momentum may lose steam. Expanded military expenditures and the challenges of an increasingly entrepreneurially driven twenty-first-century environment may slow future space progress. In addition, China and India, among others, face a staggering agenda of past and future problems, including the simultaneous needs to support hundreds of millions of people who live at or below the poverty line, deal with emerging environmental stresses on their economies, and manage the rising demands of those sectors of their populations exposed to the growth economy who expect continued prosperity. This raises questions about whether Asian countries will be able to sustain the high costs required for space leadership over the long term.

Other questions that make the future difficult to predict relate to security concerns. Will technological pressures, the complexities of multilateral arms control, and rising nationalism overwhelm prospects for military restraint among Asia's space powers? Will China's rise and India's and Japan's efforts to balance it lead to the formation of hostile alliances in space? Or will mutual desires for prosperity among all Asian countries and associated forces of economic globalization lead Asia into a more cooperative direction in space, including with outside powers? If we are to understand (and seek to prevent) negative global implications from the movement of Asian countries into space, we must examine more carefully the domestic motivations of Asia's new space actors, the nature of their regional interactions, and the challenges and opportunities they pose for twenty-first-century space security.

CONTENTS, FOCUS, AND THEMES

This book analyzes these and other questions with the aim of offering a comprehensive overview of the emergence of Asia's space programs, their current national trajectories, and their international interactions—both cooperative and competitive. For these reasons, it focuses particular attention on the roots of existing *regional* space competition, its parameters, and factors that may influence its future (such as cost, communications, and fear of spillover into new security risks). While some readers may be uncomfortable with the phrase "Asian space race," this book does not assume some foregone conclusion about its outcome, as some level of competition can be a positive factor in pushing national progress and advancing technology. Many races also end peacefully, although not all do. What is clear now is that Asia's emerging space powers *are* keeping score and *are* closely following relative gains by their rivals. The intensity of the race increases as one moves closer to the core China-India-Japan nexus and dissipates as one moves toward more developmentally focused countries, such as Malaysia, the Philippines, Thailand, and Vietnam. These countries cannot hope to keep up with the leaders, and thus they draw on foreign technologies to deliver critical space data and services while seeking to develop certain domestic capabilities for space to serve their emerging national technological infrastructures. But other regional rivalries exist too. South Korea is struggling to compete in space for military purposes against North Korea and for economic reasons against China, Japan, and others. Taiwan faces similar challenges concerning China. Meanwhile, still other countries— such as Pakistan and North Korea—are seeking to develop means of countering their more capable rivals despite having only minimal resources available for space technology. Australia is joined at the hip to the United States in space, which involves it in monitoring key adversaries, such as China. Singapore is also concerned about its neighbors and is developing a mixed program with independent reconnaissance capabilities. Thus, Asia's space "races" might be more descriptive of what is actually going on.

But rather than simply assume Asian motives and try to abstract decision making about space with a rational-actor model and "black box" assumptions, this book takes a bottom-up approach in seeking to understand the role space activity plays in the specific national politics, cultures, and histories of Asia's major participants, including their plans for economic development and their self-perceived regional and security identities. It does so by asking a series of consistent questions of each country: Why is it becoming more active in space today? What are its motivations and relations with other countries? What kinds of capabilities is it seeking in both the civilian and military sectors? And how are the country's activities likely to affect the interests of the United States? In addition, this book seeks to understand these questions within the context of broader conceptual debates in comparative politics and international relations theory, such as the possible advantages of late development for Asian advancement in the space field, the potential problems presented by declining American hegemony, and the prospects that regional or international management mechanisms for space might be developed.

Because of this book's focus on space activity and dynamics within the Asian region, it is important to delineate the definition of "Asia" used here. Its coverage is deliberately limited to those countries whose main attentions lie within the Asian region rather than including those states along its periphery that identify more with the Middle East or are members of the former Soviet Union. This definition therefore excludes Iran, Iraq, Israel, and Turkey, whose political, economic, and military focus lies primarily with their proximate neighbors, as well as the states of Central Asia, whose lineage and industrial links to the Soviet space program put them logically into a different analytical category. For this book's purposes, Asia is defined as the region bordered by Pakistan in the west, Japan in the east, and Australia in the south. Given their comparatively advanced role in terms of technology, the book's main focus is on the space programs of Japan, China, India, and South Korea, although ten other emerging space actors will be discussed as well. The aim of the book is not to provide a definitive operational history of each of these countries in space but rather to explore the economic, political, and strategic significance of their programs and the implications of their joint rise as space participants.

Fortunately, a number of very useful individual studies and histories of various national space programs exist already. These include Saadia M. Pekkanen and Paul Kallender-Umezu's *In Defense of Japan: From the Market to the Military in Space Policy* (2010); Roger Handberg and Zhen Li's *Chinese Space Policy: A Study in Domestic and International Politics* (2007); Brian Harvey's *China's Space Program: From Conception to Manned Spaceflight* (2004) and *The Japanese and Indian Space Programmes: Two Roads Into Space* (2000); Rebecca Jimerson and Ray A. Williamson, eds., *Space and Military Power in East Asia: The Challenges and Opportunity of Dual-Purpose Space Technologies* (2000); and Joan Johnson-Freese's *The*

Chinese Space Program: A Mystery Within a Maze (1998) and *Over the Pacific: Japanese Space Policy Into the Twenty-first Century* (1993). Highly informative articles, book chapters, and significant reports by other Asian space experts also provide excellent information and analysis on national space developments in Asia, including those writing on Japan (Setsuko Aoki and Kazuto Suzuki), China (Dean Cheng, Gregory Kulacki, James Lewis, Jeffrey Lewis, Kevin Pollpeter, and Larry Wortzel), India (Dipankar Banerjee, Kartik Bommakanti, Rajeev Lochan, Dinshaw Mistry, and Sundara Vadlamudi), South Korea (Changdon Kee, Kyung-Min Kim, and Daniel Pinkston), and Australia (Brett Biddington). Many of these authors are quoted from later in this book. However, the gap that this book seeks to fill is to provide a single volume covering the *comparative* motivations for space activity among Asian countries and the regionwide context. The one prior study that comes closest to this goal is K. K. Nair's useful book *Space: The Frontiers of Modern Defence* (2006), but his coverage is limited mostly to the military sector. Thus, this study attempts to capture the internal dynamics of fourteen Asian space programs across the spectrum of their activities *and* the trends seen in their international relations. In doing so, it seeks to show how national priorities affect the regional level and how the regional level could influence international stability in regard to space.

To accomplish these tasks, this book is organized into seven chapters. Chapter 1 focuses on the conceptual themes of late development and regional competition. It explains how the relative availability of space technology in the early twenty-first century, the speed of recent space developments, and the failure of corresponding political institutions place Asia in an unstable position regarding space security. This chapter also observes that regional factors and rivalries must be taken into greater account if the international community is going to understand and interact effectively with Asian powers concerning space. In the United States, a major point is that Asian space activity and related message sending is not "all about us."

The next five chapters provide political histories of the leading Asian space programs, focusing on (1) the role of space in the particular society; (2) its economic, military, and political motivations for space activity; and (3) its future plans. Chapter 2 analyzes the Japanese space program, the oldest program in Asia and the one with the closest ties to the United States. A major focal point is the significance of the 2008 Diet Law allowing

Japanese military uses of space for the first time. This marks a major shift in Japanese attitudes toward space and has set off debates about the program's future direction within Japan itself. China is the focus of chapter 3, which analyzes its space program's slow rise from a poor stepchild of the 1960s-era missile program to a dynamic, centralized, and well-funded national effort that now strikes fear into U.S. military and civil[16] space planners. Key questions to analyze are the role of the People's Liberation Army versus the emerging commercial sector and whether China's rhetoric about space cooperation will lead to meaningful military restraint and the fostering of stable space cooperation with foreign countries, including the United States. Chapter 4 delves into the Indian space program, whose history has followed a trajectory similar in some ways to Japan's, involving significant early assistance from the United States and, to a lesser extent, the Soviet Union. India developed an almost exclusively civilian-oriented program early on, focusing especially on space applications to benefit its large and dispersed population. But India lost access to U.S. launch technology in the years following its 1974 nuclear-weapon test and then pursued a largely autonomous development program, rather than becoming dependent on Moscow. More recently, India has begun to conduct military space activities and put more funding into high-prestige exploratory missions to the Moon, showing a clear desire not to be eclipsed by China. Chapter 5 describes and analyzes South Korea's remarkable rise as a space actor. Although the Republic of Korea lacked any significant space capability fifteen years ago, it now boasts sophisticated remote-sensing capabilities and nearly state-of-the-art production facilities for satellites. While it is still experiencing growing pains in the space-launch field, it has the potential to move forward rapidly if national budgetary support remains intact and stable. Chapter 6 offers summaries and analyses of ten emerging space powers that are beginning to play a significant regional or international role and are likely to expand their space capabilities in the future: Australia, Indonesia, Malaysia, North Korea, Pakistan, the Philippines, Singapore, Taiwan, Thailand, and Vietnam.

Finally, chapter 7 draws on the information and themes discussed in the country profiles to assess regional trends and to consider possible global implications. Clearly, if Asia's space race turns primarily in a military direction, there will be serious regional and international implications. Yet, even short of military conflict, many observers believe that Asia's dynamic entrance into space activity challenges future U.S., Russian, and European leadership in space. This raises questions about how the international community might best respond. If current space powers have worked with Asia's emerging space actors to elaborate and construct a system for space management and joint development, possible negative implications could be greatly reduced. If, however, space remains reliant on existing, largely cold war–era agreements and if countries fail to *prevent* the emergence of norms accepting of space weapons and conflict, then regional rivalries, U.S.-Chinese disagreements, and perhaps broader space instability could become major international problems in the future, possibly spilling over into warfare. Indeed, the character of Asia's space relations and the success or failure of these countries in managing their existing space rivalries and conflicts with outside powers may well determine the future of space security more generally—whether confrontational or cooperative—with critical implications for global security relations in the twenty-first century.

CHAPTER ONE

ASIAN SPACE DEVELOPMENTS

Motivations and Trends

Much has been written about the dynamics of the cold war in space, what journalist William Burrows calls the "first space age."[1] The United States and the Soviet Union dominated space and spent billions on sophisticated military systems to monitor the Earth from orbit. Yet neither deployed significant, dedicated weapons in this new environment after an initial foray of nuclear weapons testing in space from 1958 to 1962, which rendered a number of their own orbiting satellites inoperable. Despite a series of Soviet antisatellite (ASAT) tests from 1968 to 1982 and one U.S. ASAT test in 1985, a surprising trend of strategic restraint in space prevailed throughout the cold war, thanks to the evolution of bilateral norms, treaties, and regularized contacts on space security matters.[2]

The second space age arguably began with the emergence of significant Chinese human spaceflight capabilities in October 2003. After the lull in space competition that followed the Soviet break-up in 1991, space once again became a competitive environment. It now promises to be the focus of increasing great-power attention in the twenty-first century, across the range of civilian and military activities. But the dynamics of today's space competition differ from those of the U.S.-Soviet space race during the late twentieth century, given changes in the world and differences in the emerging actors and their relationship to space technology. This context includes some factors that favor greater competition and others that might support increased space cooperation.

First, the international system is now characterized by multiple great powers, not by bipolarity. This is a far cry from the us-versus-them world of the superpowers, whose leaders did not have to worry about the activities of third countries in space. Despite their political hostility to each other, these conditions made cold war space management and the development of consensual norms much simpler. The higher "transactions costs" required to craft and enforce *multilateral* agreements make space management today arguably much more difficult. This problem is exacerbated by the widely disparate perspectives on space security seen in Asia today, unlike the relative consensus that prevailed between Washington and Moscow in past decades and the high levels of agreement seen within modern Europe.

Second, and related, emerging military space threats are not linked to any arms-control process or some similarly iterated security negotiations among the critical actors that might promote mutual restraint. Instead, the bulk of new countries in space lack a history of participation in arms control and have had no such talks involving space security with their rivals. Most of those in Asia share several common characteristics, in that they are relative newcomers to military space activity, are still engaged in longstanding regional competitions, and have no history of discussing these sensitive matters with their neighbors (or with the more established space powers). Formal negotiations on space at the Conference on Disarmament (CD) in Geneva were blocked during the 1998–2008 period by a feud between the United States and China. Washington had other, higher priorities it sought to pursue at the CD, such as negotiating a Fissile Material Cut-Off Treaty (FMCT), and did not view talks on space as either urgent or particularly desirable, given its interest in keeping its options open for possible space-based missile defenses. Beijing did not want to give up its ability to produce fissile material while it believed the United States was pursuing unilateral advantages in space. The U.S. position on the desirability of space security talks changed after the 2007 Chinese ASAT test and the 2008 election of President Barack Obama, and the Chinese government now supports FMCT negotiations. However, the promised initiation of CD talks on space has been blocked by Pakistan, because of its security concerns vis-à-vis its larger rival India. Thus, formal space security talks remain in abeyance, and gaps in the existing arms-control framework—that allow testing and deployment of kinetic space weapons,

lasers, other directed-energy systems, and dual-capable missile defenses—
continue to cloud the future of space security.

However, a third factor that sets today's space context apart from the
cold war is working in the other direction. Recent economic globalization
has spread information, technology, and investment around the world in
such a way that there are many more cooperative pressures affecting space
activity than ever before. The typical space corporation is multinational
(from satellite providers such as Intelsat to launch services providers such
as International Launch Services), using technologies from more than one
country and marketing their products and services worldwide. Even political rivals (such as the United States and China) now trade extensively,
which provides an incentive for space cooperation, although little has yet
occurred in the U.S.-Chinese space context since 1998. In addition, growing financial interdependence and international trade play an increasing
role in state decision making, affecting not just moderate-sized countries
but also the world's previously greatest creditor, the United States. China's
possession of large denominations of U.S. debt and the two sides' yearly
buying and selling of half a trillion dollars of each other's products make
the two sides highly dependent on each other, in sharp contrast to the economic autonomy of the cold war superpowers. The tremendous economic
value of these exchanges significantly raises the costs to both sides of any
future space conflict.

Finally, a fourth emerging (and positive) factor is that there is more
widely spread scientific knowledge after fifty years of space activity about
problems caused by the release of harmful debris and electromagnetic
pulse radiation into the orbital environment.[3] Despite China's 2007 kinetic
ASAT test, this factor—at least in theory—should promote greater international cooperation in trying to *prevent* collective damage to space. There
has indeed been some progress in this direction, such as the passage of a
set of Orbital Debris Mitigation Guidelines at the United Nations in December 2007. The problem is that these guidelines lack an enforcement
mechanism and remain voluntary. At the same time, the dramatic increase
of the *value* of space activity to the societies, economies, and militaries of
the world in recent decades suggests that self-interest alone should promote
future restraint, at least among major spacefaring countries.

Asia's place in this emerging set of space-related challenges is significant. Besides the United States, Russia, and the countries that make up the

European Space Agency, almost all of the most rapidly developing space programs are located in the region. China, India, Japan, and South Korea are leading Asia into space, but countries such as Australia, Indonesia, Malaysia, North Korea, Pakistan, the Philippines, Singapore, Taiwan, Thailand, and Vietnam all have significant space plans of their own. For this reason, understanding the dynamics within and among Asia's emerging space programs is critical to coming to terms with the second space age.

The rest of this chapter first takes a brief look at lessons from the cold war in space to set a foundation for possible Asian learning from this experience. It then turns to the problem of understanding the new challenges we face in moving from bipolarity to multipolarity while adding new actors with unresolved histories of hostile relations (and without the rapprochement seen, for example, in the post-1945 Franco-German case). The chapter next considers some of the motivations of Asian actors in pursuing space capabilities and their greater options as "second-generation" states in space. Finally, it attempts to place Asia's twenty-first-century space competition into the context of broader trends in Asia's emerging international relations. Current predictions are generally pessimistic about hopes for collective management. These arguments, however, set out a challenge to national decision makers and provide a rationale for studying the individual histories and perspectives of the various Asian space programs, which are analyzed in chapters 2 to 6.

LESSONS OF THE COLD WAR IN SPACE

Space relations among Asian countries today differ significantly from the norms of the cold war in space. Despite hostile political relations between Moscow and Washington, the cold war witnessed the emergence of a regularized set of interactions aimed at managing the possible harmful effects of space competition, especially in the military arena. In fact, a culture of "managing" space activities began early on in the cold war context, with space's inclusion in the Partial Test Ban Treaty in 1963 (which halted U.S. and Soviet nuclear testing in space), the negotiation of the 1967 Outer Space Treaty (which banned weapons of mass destruction [WMD] from orbit, demilitarized the Moon, and forbade the claiming of territory on any celestial body), and in the arms-control negotiations leading up to the 1972

Strategic Arms Limitations Treaty and the Anti-Ballistic Missile Treaty. These latter two agreements formalized an existing de facto U.S.-Soviet norm banning interference with each other's treaty-verification and other information-gathering satellites and went further to codify a prohibition on the development, testing, or deployment of space-based ballistic-missile defenses. While a post–cold war U.S. decision terminated the ABM Treaty in 2002, the agreement served during the early 1970s to prevent a worsening of the U.S.-Soviet technological arms race, which very likely would have developed an active space-based component as well. Additional U.S.-Soviet leadership helped forge supporting international norms and legal requirements in civil space: the 1968 Rescue and Return Agreement for astronauts in distress, the 1972 Convention on International Liability for Damage Caused by Space Objects, and the 1975 Convention on the Registration of Objects Launched into Outer Space. Subsequent U.S.-Soviet cooperation in the Apollo-Soyuz project in July 1975 and engagement (albeit without success) in discussions aimed at banning ASAT weapons in the late 1970s provided ongoing contacts over space security. But détente relations were not a prerequisite. Space security negotiations continued even during the height of U.S.-Soviet political disputes during the Reagan administration in the early to mid-1980s, because of the perceived risks involved of *not* talking about space.[4] Thus, while hostile relations continued relatively unabated throughout the cold war, a norm of managing space *bilaterally* emerged. Regular contacts and exchanges regarding critical issues of concern remained the rule rather than exception. Such communications, contacts, and norms of consultation have been almost wholly absent from Asian space relations. As each country accelerates its military space program, the absence of such conflict-prevention mechanisms for space could become a serious problem.

In sum, the benefits of long-term communication regarding space between the two capitals allowed the formation of both a series of formal and informal agreements over time in space that created at least a limited framework for stability. Since the two states feared the possible spillover of space conflict into general war, they decided to form preventive measures to avoid destabilizing deployments in space and to create networks for reliable communications and periodic cooperation. Since the end of the cold war, the United States and the Russian Federation—which are still the world's two most capable space programs—have cooperated significantly in the area of space technology. Russian engines power American Atlas rockets for launching U.S. military payloads, and Russian Soyuz spacecraft routinely deliver American astronauts to the *International Space Station* (*ISS*). But this culture has not been adopted by the latecomers in Asia, for a number of geostrategic, economic, and political reasons. This lack of Asian cooperation in space security—and in other areas of space activity as well—poses a risk of future failure in space conflict avoidance and crisis management. Indeed, disputes over space activity could quickly spill over into military disputes and put space itself at risk if countries react hastily in a crisis and lash out against the space assets of one of their rivals or those of an outside power, possibly creating dangerous orbital debris or—in a nuclear scenario—disabling perhaps hundreds of spacecraft with electromagnetic pulse radiation.

Historically, the political hostilities spurred during technological races have often led to an increased risk of conflict, as countries fail to consider alternative courses of action and feel "forced" to react to the behavior (or predicted behavior) of others in the race.[5] As the military analyst and China expert Larry Wortzel observes: "Whether one is a proponent of arms control agreements or not, the dialogue between the U.S. and the Soviet Union over arms control and treaties produced a body of mutual understanding that holds up today."[6] Norms and rules emerged from this process that ruled out extending sovereignty into space and allowed safe passage of each other's satellites. However, Wortzel points out, "No such dialogue has taken place with China."[7] To date, both sides are to blame: the United States pursued a strategy of unilateralism regarding space security for a decade, and China has to date rebuffed direct military-to-military talks on space. Increasingly, experts on both sides identify the lack of such a dialogue as a worrisome problem. Whether the Obama administration's recent openness to dialogue on space security at the UN CD will help to solve this problem remains to be seen. It may be fruitless if the People's Liberation Army is not involved or is not open to potential limits on its space behavior. High-level political buy-in on both sides will be required for new understandings and new norms to be reached.

OLD THINKING IN A NEW INTERNATIONAL SYSTEM

In the U.S. debate on the future of space, many commentators, officials, and members of Congress alike have since 2007 simply inserted China into the Soviet Union's old slot and returned to viewing space from a more or less cold war perspective.[8] But this mode of thinking may be based on false assumptions that are distorting the U.S. understanding of current dynamics in space. Moreover, China—as will be discussed in chapter 3—is *not* the Soviet Union and instead has unique political, economic, and security interests that must be understood and taken into account. A related problem is that the United States has been used to seeing itself as the center of attention for all other nations since 1991. Without question, during the cold war, Moscow and Washington watched each other's every move carefully and tended to react to each other's programs in a more or less symmetric fashion. Today, however, the new players in space have considerably less capability to match the United States mission for mission or technology for technology. The world is also multipolar in structure, and they have other regional actors to worry about who may be *more* important to them, in terms of their national goals and interests, than Washington.

China, for one, is very sensitive to developments in Asia. While China may be competing with the United States in space, it is equally interested in its relative place with respect to its Asian neighbors. Moreover, when China sprints forward in its space activity, there is no question that India, Japan, and South Korea all feel challenged and want to react. China's manned spaceflights and military tests have caused significant concern in New Delhi and Tokyo, in particular. Thus, while U.S. behavior in space does matter and is followed carefully by Asian space actors, "neighborhood questions" are also an important factor for understanding current Asian space motivations. This is good and bad. It may allow Washington to influence the activities of Asian countries through its own example and its potential ability to draw in countries for cooperative projects. But it may reduce Washington's leverage in seeking to "manage" Asia's space competition in cases where their priorities lie more in pursuing specific advantages versus their neighbors. A central (and difficult) question is how to determine what cases are likely to receive which response.

Observers of foreign motivations in space activity—who frequently lack information about related domestic politics—often use simplifying principles such as self-interest, rationality, or mirror imaging in their efforts to interpret policies by Asian governments. In many cases, these efforts distort the actual reasons for decisions, which often emerge from complex bargaining, compromises, power grabs, experiments, or even misunderstandings within national systems or between rival technological and military bureaucracies. Instead, it is too often posited—by analysts simply assuming a unitary rational-actor decision-making model in the absence of data—that authoritarian countries like China always have a clear and insidious "plan." As the U.S. analyst John Tkacik argues, "Beijing's political and military leaders alike foresee 'competition' in space with the United States. They certainly plan to seize the high ground of low-Earth orbit and then will likely move to . . . even higher ground."[9] But it is not clear what actual plan the author is referring to, who approved it, how it stands to be implemented in the future, and, indeed, whether it exists at all. As Roger Handberg and Zhen Li argue: "Authoritarian governments often appear more monolithic than they are in fact. But, their politics are often a constantly changing combination of personalities and factional politics."[10]

A common fallacy in political analysis is to assume that other countries have no equivalent of the various obstacles to coherent policy found in one's own domestic politics. Analysts should not rule out this possibility in viewing Asian space developments, even controlling for the seemingly mitigating factor of authoritarian politics in a number of Asian countries. As the political scientist Robert Jervis reminds us, "The fact that people must reach decisions in the face of the burdens of multiple goals and highly ambiguous information means that policies are often contradictory, incoherent, and badly suited to the information at hand."[11] Equally authoritarian politics in the Soviet Union often seemed "enviable" to American analysts during the cold war, since the leadership could apparently get what it wanted. Yet Moscow made many mistakes. Soviet military officials, for example, put undue stress on the allegedly "evil intentions" behind certain U.S. space plans, overlooking the role of domestic politics in the U.S. decision to pursue the space shuttle in the 1970s. They then convinced their political leaders to fund costly research to build a Soviet military equivalent, the *Buran*. The program eventually spent the equivalent of billions of dollars against a threat that did not exist, wasting precious resources and eventually helping to bankrupt the Soviet government. Thus, while space decision-making processes may seem rational and clearly directed from an external viewpoint, the reality may be more complicated and the message not at all what some observers expect.[12] That said, we should not rule out the possibility of underreaction to other countries' space policies either. There are historical cases when countries have failed to prepare for emerging threats. The difference in space is that there is greater transparency in terms of actual testing of new technologies and potential weapons, which may eventually promote a greater clarity of purpose.

In order to understand Asian space politics better, it is more important to understand how Asian space programs see themselves and their surroundings, including other countries, than it is to simply ask, "what would I do?" As the former U.S. State Department official James A. Lewis argues, "A mirror-image model could distort our understanding of Chinese programs."[13] In other words, we need to grasp, in Jervis's terms, "how decisionmakers perceive others' behavior and form judgments about their intentions."[14] Similarly, given the multipolar context in Asia today, all capitals need to be cautious not to overreact and not to adopt an assumption that its own country is the bulls-eye for policy decisions by all others. Not surprisingly, the political science literature on decision making has long identified a common tendency in "the propensity of actors to see themselves as central to others' behavior."[15] While there are examples from the history of Asia, we do not have to look far for evidence of this tendency in recent U.S. national security policies.

For example, the United States intelligence community produced a document in 2004 that came to be known as the Duelfer Report.[16] This exhaustive study aimed at determining the facts behind Iraq's WMD programs, benefiting from the unique postwar opportunity to conduct hundreds of interviews with former Iraqi officials and scientists. Its conclusion surprised many U.S. observers. Contrary to U.S. assumptions prior to the 2003 war, it turned out that Iraq had halted all of its WMD-reconstitution efforts except in the chemical weapons area, where it hoped to maintain a capability against Iran.[17] By making mistaken assumptions about Iraqi intentions and their worldview, the United States made a major miscalculation and has spent much blood and treasure over the past decade. In space, these pitfalls need to be avoided if countries are going to prevent unnecessary conflicts, safeguard their limited budgetary resources, and prevent a perhaps unnecessary cycle of action-reaction arming for space. As experts know, space conflict will likely involve considerable collateral damage to other orbital assets, given the physics of space, and U.S. ownership of a majority of currently operating satellites puts it at greatest risk. As the military analyst John Arquilla argues, "A major war in space would likely harm the United States far more than other nations."[18] Whether preventive regional diplomacy or other means might reduce these threats will be explored in this book's conclusion.

Despite these cautions, much Western analysis may already be making some of the same mistakes concerning Asian space activities. As Gregory Kulacki and Jeffrey Lewis observe regarding the January 2007 Chinese ASAT test: "Most American analysts placed the United States at the center of Chinese calculations, either by asserting that the test was part of a deliberate effort to acquire a comprehensive set of counterspace capabilities to offset U.S. military superiority, or an attempt to induce the United States into negotiations governing uses of outer space."[19] Instead, these experts argue, based on some eighty interviews conducted in China, that "the decision to flight-test the hit-to-kill interceptor was not determined by any external event or series of events, but by the maturity of the technology." That is, rather than being a specific "message" to the George W. Bush administration, the test was a "technological demonstration" that marked the culmination of a long-term research and engineering effort that had begun in the 1980s in response to the U.S. Strategic Defense Initiative and to U.S. ASAT efforts at the time.[20] By 2007, any immediate political intent may have been lost. This research does not change the risk that the new Chinese capability poses to U.S. assets, but it does highlight a possible gap between China's technical and political bureaucracies in regard to space. At the very least, these alternatives to the unitary-actor model merit closer investigation than have been provided to date by mainstream analysis. While China may have a plan for space, it may not be as clear as observers think it is, and it may not be set in stone, particularly if the country's political, commercial, and military actors are not on the same page.

With these cautions in mind, this book seeks to uncover, understand, and explain the national motivations and sociopolitical contexts driving each Asian nation's emerging space program. In order to address these issues, the analysis employs a multitiered framework beginning with domestic perspectives and national priorities for space, then moving to analyze regionwide interactions and trends, and finally considering implications at the international level. At the domestic level, it begins by introducing the economic concept of "late development" as a shared factor in influencing and shaping the behavior of Asia's space actors. While each country has sought different partners and followed different paths, the ready availability of space technology on the international market has clearly spurred the pace of Asian countries' recent progress in space. Yet the relative ease of technological acquisition may also have exacerbated prospects for the successful management of intraregional conflicts related to this new environment by creating false hopes of national prestige, dominance, and influence. Such technological "pacing" may have encouraged greater space management during the cold war. Similarly, space is more closely tied to both nation building and national economic development in Asia than it was for the two superpowers. Notably, regional trends in Asian space activity today point to a dominant trend of what Joan Johnson-Freese calls "techno-nationalism,"[21] which can be seen in the patterns of competition and cooperation within the region and in the tendency of all of the major Asian powers to seek technological autonomy rather than the cost savings to be achieved through a division of labor with other spacefaring nations. Such dynamics have made the accomplishment of any significant cooperative framework for Asian space activity politically impossible to date, in sharp contrast to the highly integrated programs of the European Space Agency.

These trends also characterize Asian security relations more generally. Thus, it is important to situate space within a broader set of competitive trends that are emerging in Asia, particularly regarding the problem of regional order. This chapter's final section will address these points. Whether space will play a positive role in helping to prevent or manage future conflicts will be an important element in twenty-first-century Asian and global stability. If regional rivals fail to rein in tendencies toward conflict, space may become an area where an arms race emerges. If, on the other hand, countries in the region can rise above current national differences and identify common interests in the management of space and the use of space to promote regional development and environmental stewardship, space may instead become a valuable tool for improving Asia's international relations.

ASIA'S SPACE PLAYERS AS "LATE DEVELOPERS"

An increasing proportion of the economies, societies, and defense establishments of modern states rely on information from space, giving all countries an increasing interest in being able to access space independently. Yet *when* countries have decided to move into space and *how* they have decided to do it—in terms of direction, technology, and resource commitment—have differed substantially over time. What makes space developments in Asia interesting is that the region is witnessing the simultaneous acceleration of space activities among many of its major powers. What explains this recent interest, and what factors are likely to shape these developments? Why has this development been especially rapid of late, and what are the possible implications of the lack of integration of these countries into existing norms about space security?

The practical process through which Asian countries have entered space can be linked to broader historical trends involving the transfer of technology and know-how from more developed to lesser-developed countries. A number of general findings from the literature on these processes are relevant for understanding the emergence of Asia's space race. In the early twentieth century, the work of the American political economist Thorstein Veblen analyzed the rise of imperial Germany and its success in using foreign-origin technology (such as the railroad) to surge past traditional industrial powers (such as Britain) in the application of these instruments. As Veblen argued in 1915, "as a matter of history, technological innovations . . . have in many cases reached their fullest serviceability only at the hands of other communities and other peoples than those to whom these cultural elements owed their origin and initial success."[22] The logic of Veblen's argument emphasizes the ability of latecomers to start at a higher level of technological development, without the "baggage" of the old system of cultural organization, including prior forms of thinking, production, and distribution. As an example, Veblen refers to the advantages accrued by the U.S. Steel Corporation in the late 1800s, whose development from the outset could focus on the application of more advanced technology, a larger scale of production, and a wider network of rail-based distribution than those in Europe who first developed the steel industry.[23]

In space, it is clear that Asian countries have benefited from several factors of late development: (1) the availability of foreign technology; (2) the know-how that certain space activities are possible and can best be achieved through certain known and tested means; and (3) the ability to channel funding into areas where space technology offers the greatest advantages (such as communications, reconnaissance, navigation, meteorology, and certain areas of space defense), rather than wasting precious resources on costly schemes that have failed in the past. Arquilla compares the U.S. role in space today to the British role in guaranteeing freedom of the seas in the nineteenth century, observing: "The U.S.-created global positioning system (GPS) . . . is having profound effects on commerce, industry, and even agriculture."[24] But, while U.S. companies have benefited by creating a wide variety of GPS-service industries, Arquilla cautions that this "public good" gives second-generation space adversaries the possibility to use the GPS system to guide their own weapons in a military conflict, particularly given the strong economic pressures on the United States not to shut the system off.[25] Drawing on the U.S. example, however, several countries (including China) are now building their own GPS-type systems to ensure continued access to these signals and the future independence of their space infrastructures.

Another benefit to latecomers is that space science and engineering are now well-established disciplines in the universities of major space powers, where latecomers can send their students for advanced training. Many of these fields and technologies did not exist for the first-generation space powers, that is, the United States and the Soviet Union. In terms of emerging technologies, latecomers can often skip first-generation systems and purchase or develop more advanced capabilities, drawing on the knowledge of what has come before them. U.S. Air Force Major General James Armor explains these advantages from a military perspective:

> Foreign states, unlike the U.S., are not tied to legacy systems with architectural components that are twenty years behind the state of the art. Instead, they are capitalizing on the latest small satellite and sensor designs, and responsive launch system technology, and are employing the latest computers to process and integrate their data into higher fidelity products at a fraction of the cost.[26]

In the 1950s, the Russian-born economist Alexander Gerschenkron further developed concepts of technological borrowing by examining the *political* implications of late development. Specifically, by looking at the Soviet experience, he argued that the recognized need to "catch up" with the West in terms of technology strengthened forces of political centralization and helped make authoritarianism more likely among successful late developers, including in Asia.[27] For these reasons, he saw the rush toward development as creating problems for both political stability and international security, especially when drives to develop technology fomented chauvinist versions of nationalism. Gerschenkron concluded from these conditions: "The paramount lesson of the twentieth century is that the problems of backward nations are not exclusively their own. They are just as much problems of the advanced countries."[28] Indeed, looking back on the experience of the early to mid-twentieth century, Gerschenkron concluded, "The road may lead from backwardness to dictatorship and from dictatorship to war."[29] In looking at Asian space development today, one may ask whether the possible emergence of a hostile space race represents a similar danger in the absence of cooperative mechanisms to avoid conflict.

THE EMERGENCE OF ASIA'S NEW SPACE POWERS

Historically, the motivations for countries to develop space programs have tended to focus on three distinct but interrelated goals: scientific-technological progress, national security, and international prestige. All of these features can be seen to a greater or lesser extent in Asia's space programs today. But how this package of incentives has been defined, shaped, and implemented has differed in each country. The technological challenges they face are similar, although when countries have pursued them, how they have organized themselves, and what areas have become priorities have varied.[30] In certain respects, major space programs share analogous factors with nuclear programs. Like nuclear efforts, space programs are very expensive to develop and carry out. Unlike nuclear weapons, however, they do not bring a perceived guarantee of security, which partly explains their relatively later development. Instead, they have historically brought economic and technological benefits, information on rivals' military programs through passive reconnaissance, and international prestige, categories more typically associated with the concept of soft power than with hard power.[31] This is not to say that there are not hard-power uses for space: the United States has used space for force enhancement and precision guidance since the 1990s. But space itself has not been a location of conflict. Whether this trend will continue in the twenty-first century remains to be seen.

Scientific-Technical Motivations

One of the shared factors behind Asia's current space programs has been a desire for scientific-technical progress and its related economic benefits. A frequent early goal of second-generation space programs is gaining access to foreign-generated space data through a ground station or other downlink technology. The second step is typically the lease or acquisition of a foreign-built satellite to gain experience in command-and-control operations from the Earth while using the satellite to gather data specific to their country's needs.[32] The next steps are learning how to build satellites domestically and, if there is adequate funding and available technology, gaining the ability to launch them. Such efforts may be driven by top-down pressure from politicians and government officials or bottom-up pressure from scientists, space enthusiasts, and entrepreneurs, who are either fascinated with space for research purposes or motivated by the possibility of providing space-derived services to the general public. In both routes, the pursuit of space technologies requires large-scale educational, infrastructure-related, economic, and even social changes in order to sustain them effectively. They also demand a long-term budget commitment from national governments, because space programs require multiyear timescales and plans that extend beyond the terms of most presidential administrations, even in authoritarian countries. In times of financial austerity, space expenditures may be seen as a costly luxury, leading reformist politicians to call for cuts or even the elimination of whole projects. But public fascination with space technology, its risks, and its benefits may help sustain budgetary support through tough times.

Space activity in Asia today is commonly identified with notions of scientific-technological progress, "modernity," and the promise of gaining new economic and social benefits from the application of advanced space technologies to current problems. Space thus plays a symbolic role in Asia, much like the civilian acquisition of nuclear technology did in the 1970s and 1980s, which showed nations' commitments to take on such challenges as energy generation and health care concerns (via the use of radioisotopes for cancer treatment). These programs also pushed forward their national scientific and economic infrastructures and provided at least latent security benefits (through the possession of nuclear fuel).

Much of the initial value of space activity has been associated with information-technology advancement and the regional prestige gained from orbital accomplishments. It is notable that space technology has frequently been linked by Asian governments to making developmental progress in such fields as agriculture, forestry, city planning, communications, disaster warning, education, fisheries management, telemedicine, transportation, and water conservation. Asian countries have historically focused less attention on space exploration, in part because there have been few "firsts" in near-Earth space that have not already been accomplished by the two superpowers and in part because they have had fewer resources available for such missions. But this trend is now changing. Notably, the Soviet and American space programs witnessed early military uses of space. By contrast, most of the programs in Asia (with the exception of China) began with civilian applications and infrastructure building and have only recently turned to military uses. It remains to be seen if these peaceful legacies will have any constraining effect on the future development of Asia's space programs or whether their publics and governments will willingly accept significant shifts in public expenditures toward expensive national security space technologies.

National Security Motivations

Space assets can provide uniquely useful information about adversaries and improve the effectiveness of existing weapons through the advantages of precision guidance. To date, only the United States has made extensive use of precision-guidance technology for military purposes, in part thanks to its development and ownership of GPS technology (whose lineage stems from U.S. Navy Transit and Navstar satellites of the 1960s to early 1980s).

In terms of the more widely dispersed technology of satellite-based reconnaissance, space has provided a "high ground" from which Asian countries can view their military rivals in the region. For those with more pressing national security concerns, such as for China in regard to Taiwan, space weapons may be identified as asymmetric tools to prevent possible intervention by the United States or others. Individual space weapons do not offer the kind of revolutionary destructive power of nuclear weapons. But they may be seen as potentially effective in bolstering deterrence and in reducing adversary confidence in the ability to conduct space-supported military operations against targets in the region.

Space launchers, of course, can serve as WMD delivery systems, although the technologies optimal for space launch (liquid-fuel boosters) have drawbacks as ballistic missiles because of their long and often transparent "tanking" times in preparation for launch, their typical use of fixed launch sites, and the impossibility of storing them once supercooled cryogenic fuels have been inserted into the vehicle. Still, space launchers provide at least a residual ballistic-missile capability and certainly considerable experience for national scientists and technicians that can potentially be applied to offensive military purposes.

In the mid-1980s, an article by the political scientist Raju G. C. Thomas argued that state motivations in what he called the "nuclear-missile" sector could be characterized as falling into three categories in terms of motivations: defense, development, and defense *and* development.[33] Notably, he placed one Asian country in the defense-oriented category (Pakistan) and three (India, Taiwan, and South Korea) in the category of mixed defense/developmental motivations. He omitted Japan altogether, presumably because of its perceived pacifist orientation. In terms of our understanding about space and national security motivations, it is worth analyzing how his predictions have held up.

With the benefit of hindsight, we can see some validation but also some important shifts in regard to the space-missile sector in Asia and its relationship to the nuclear-delivery role. First, Thomas wrote that "South Korea and Taiwan perceive themselves as threatened by China and the Soviet Union."[34] This helped explain their original security motivation for seeking both nuclear weapons and ballistic missiles. Today, the Soviet Union has long been broken up and the Chinese government has changed substantially, now constituting South Korea's leading trade partner. Taiwan and China have also engaged in a significant rapprochement, although not on the same scale. The combination of strong U.S. opposition and a shifting security environment eventually caused both Seoul and Taipei to drop their ambitions for nuclear weapons. But at least South Korea continues to seek a space-launch capability, and both it and Taiwan have sought to develop missile defenses, a motivation for missile acquisition not discussed by Thomas. Here, there has been an important shift in military motivations away from WMD delivery and instead toward the accomplishment of less offensively oriented military benefits, such as intelligence and defense. Taiwan has shelved its prior WMD force-projection aims and has focused instead on developing a less-threatening means of securing itself against China. But it has fallen somewhat behind its East Asian rivals in terms of space capabilities, from space launch to satellite development. However, thanks to current market conditions, it can purchase space services on the international market at a fraction of the cost of developing them itself. Seoul has dropped its force-projection aims against Russia, although not necessarily against North Korea. It still seeks conventionally armed missiles for use on the peninsula. More significantly, Seoul is strongly motivated by new commercial factors and a drive for technological advancement. South Korea is increasingly seeking to export satellites and looks forward to eventually selling launch services. Thus, while its motivation remains mixed, its military aims have changed (and declined) and its civilian aims have expanded.

By contrast, Japan has moved to adopt new military aims in space to supplement its previously exclusive civil and scientific aims, consistent with 2008 changes in the Diet law governing Japan's space activities. While it lacks any offensive orientation, it has begun to work on devoted military support assets, which may in the future assist its missile-defense program. In the commercial sector, Tokyo also seeks markets for its launch services and its satellites while maintaining a robust human spaceflight program and scientific research effort.

In regard to India and Pakistan, their testing of nuclear weapons in 1998 and deployment of significant numbers of systems, along with extensive missile testing, validate the traditional security motivation. However, there is less evidence of use of space launchers as a "cover" for military purposes than Thomas predicted. India, until recently, has kept its space and ballistic-missile programs separate and used different technologies.

Today, analysts such as Kartik Bommakanti reject the notion that either India's Geosynchronous Space Launch Vehicle or the earlier Polar Satellite Launch Vehicle could be easily adapted for ICBM use, given their unwieldy launch configurations (unsuitable for road-mobile deployment), use of certain cryogenic liquid fuels, and the absence of associated warhead and re-entry technologies.[35] Bommakanti also argues that "there are no strategic motivations for India to develop an ICBM," because of the limited range (three thousand miles) required to reach targets in the territories of its main potential adversaries, Pakistan and China, and its current possession of the solid-fuel, military-purpose Agni-3 medium-range ballistic missile. Thus, commercial and scientific motivations seem to have outweighed traditional military drivers, at least in terms of its space program. India has recently begun to move actively into the area of satellite reconnaissance, in part with assistance from Israel. Where Indian space, missile, and weapons-related motivations may eventually converge is in the area of missile defense.

By contrast, Pakistan has thus far failed to develop either a significant commercial or scientific space capability and still lacks a space-launch vehicle. While it has cooperated with China on some small-scale satellite-development projects, it has very weak space capabilities overall, drawing largely on foreign satellites as it tries to build up its domestic capability. Its purported space-launch program seems clearly dominated by its military ballistic-missile program. Thus, its motives seem to remain largely military oriented, as Thomas predicted. Pakistan's much more limited financial resources and decision to focus on nuclear weapons have also stunted its space program, compared to those of its Asian rivals.

Finally, the case of North Korea is an interesting one, although—notably—it was not mentioned by Thomas as a concern in the mid-1980s. Today, it seems clear that Pyongyang falls into his "defense" orientation category in regard to its missile aims. Since its nuclear-weapons tests in 2006 and 2009, almost all observers have concluded that its purported space-launch program is driven by a desire to acquire a long-range ballistic missile rather than any devoted space-launch capability. A key point in this assessment is North Korea's lack of any satellite capability today or related production infrastructure. Its focus has been on the rocket (or missile) rather than any civilian payload. Thus, it seems that the motivation is more closely linked to possible WMD delivery aims than for advancing its space capabilities, since it could much more cheaply purchase access to space on a foreign booster for its satellites. But analysts of domestic politics suggest that Pyongyang's desire to compete with South Korea for prestige or possible plans to develop a low-cost commercial space booster present alternative explanations to the security model (points considered in chapter 6).

International Prestige

As noted above, Asian space activity has been motivated in part by a desire for international recognition and prestige. Just as the most well-recognized technological victory of the United States during the cold war was its planting of the American flag on the Moon in 1969, China's successful launch of its first taikonaut into space in 2003 and its subsequent manned missions have had a similar regional impact because they marked firsts among Asian countries. Space has allowed China to beat out traditionally more advanced technological powers such as Japan and South Korea, as well as its rival India, in looking more "modern" and "developed." India's recent move from its historical focus on Earth applications into the field of space science and exploration, including its highly publicized *Chadray-aan-1* lunar orbiter in 2008 and 2009, can be seen as one response. New Delhi's announcement of plans for human spaceflight is another, particularly as this will involve considerable new costs and associated risks. Japan's *Kaguya* orbital mission to map the Moon with high-definition images and its *Kibo* module on the *ISS* have played a similar role, as has South Korea's astronaut So-yeon Yi, who visited the *ISS* in 2008. In addition, such missions are intended to stimulate domestic popular interest, rallying support for their respective governments, and providing incentives for young people to consider careers in science and technology.

It is this very issue of prestige that brings us back to Asia's space "race." What is clear from recent space developments is that the various participants care about *relative* regional gains. While the world's first orbital satellite, human spaceflight, and Moon landing have already occurred, the countries involved now clearly want to be the first *from Asia* or, if not, then at least first in comparison to their main regional rival. One example is the recent race on the Korean Peninsula to be the first country to orbit a satellite around the Earth successfully. While this task has already been accomplished by China, India, and Japan, such "bragging rights" have not yet been established between the two Koreas. James A. Lewis concurs: "If there is a space race [today], it is in Asia."[36] Besides the activities on the Korean Peninsula, Lewis points out that "India and Japan have announced ambitious schedules for exploration and both are expanding their military space capabilities in response to the perception that China's successes have increased its power at their expense."[37] Thus, the current Asian space race is less of a quest for world leadership in space than it is a competition for *regional* ascendancy, technological prowess, economic advantage, political influence, and ultimately power (both hard and soft). However, if it continues its current course, it could well spill over onto the global level, with possibly destabilizing effects.

Limits of Asian Space Cooperation

As noted in the introduction, one warning of potential Asian problems in space conflict prevention comes from the extremely limited evidence of space-related cooperation within the region. Similarly, there has been little effort to form mutual security approaches to space. While China, India, Japan, and South Korea have engaged in relatively extensive cooperation for the purposes of technology acquisition with more developed spacefaring powers (such as Russia, the United States, and members of the European Space Agency), they have very rarely cooperated among themselves. Moreover, they have cooperated least in the area of military- or security-related space activity. What cooperation has emerged within Asia seems largely motivated by competitive goals. The Chinese-led Asia-Pacific Space Cooperation Organization (APSCO) is a membership body established in 2008 on the basis of a less formal process of exchanges begun in 1992. It has attracted a number of minor space actors, but India, Japan, and South Korea—among others—have refrained from membership, given China's somewhat heavy hand and control over joint projects. Similarly, the Japanese-led Asia-Pacific Regional Space Agency Forum (APRSAF), a non-membership organization established in 1993, has seen little Chinese participation and only a slightly more extensive presence of Indian and South Korea scientists. Instead, Japan's focus has been on providing training and technology to lesser-developed space countries in Asia both to promote science and possibly attract future space commerce. Currently, there are no major, regionwide projects among Asia's leading space players in the scientific or commercial sectors and no significant political, diplomatic, or military talks on space among the major capitals.

One theme that will be developed further in this book is the problem of this peculiarly bifurcated structure of Asian space cooperation—that is, the presence of strong contacts with outside providers of space technology and links to weaker regional partners for the self-interested purpose of building political and economic influence. The result is a "missing middle": the absence of substantive cooperation *among* the major four Asia space programs themselves (China, India, South Korea, and Japan).

SPACE AND BROADER ASIAN SECURITY RELATIONS

Although relatively few analysts of Asian security to date have focused on its space-related aspects, there is a growing recognition that space will soon emerge as a more important part of the Asian security picture than it has in past decades. This situation is changing due to the acceleration of Asian interest in space activity in the early twenty-first century. But how might space fit into existing schools of thought that attempt to explain and predict larger Asian security trends? Since space activity will not take place in isolation from other realms of policymaking, this question requires careful examination. Thus, the final task of this chapter is to "position" space within the broader literature on regional politics, economics, and security and to consider possible interactions. The literature on Asian security, notably, is pessimistic about prospects for collective security building, at least in the near term. One concern is that space competition might *exacerbate* tensions if concerted efforts are not made in the coming years to overcome these obstacles.

Hegemonic Stability and Power-Transition Approaches

The most widely applied concepts in Asian international relations in the twenty-first century have been those of "hegemonic stability" and related theories of power transition. As Michael Mastanduno argues, "The defining feature of the new [post–cold war] international system is the dominance of the United States."[38] In Asia, he argues that Washington has "pursued a hegemonic strategy for the maintenance of order,"[39] using political, military, and economic tools. However, he cautions that with China's rise, "the prospects for order in East Asia . . . remain uncertain and problematic."[40] Since the 2007 Chinese ASAT test and evidence of Chinese military writings mentioning possible conflict in space, combined with similar writings on the U.S. side, future space relations between the two sides have come directly into this debate.

Some analysts see continuing U.S. hegemony but predict a coming confrontation because of China's unwillingness to accept limits on its behavior and desire to "rise to great power status."[41] Ashley Tellis argues that existing prosperity in the international system is due to the United States' hegemonic role and "the U.S. military's ability to police the commons."[42] But he sees U.S. satellites and space-related ground stations as its "Achilles' heel."[43] For this reason, Tellis argues that China's leaders "simply cannot permit the creation of . . . a space sanctuary because of its deleterious consequences for their particular interests" as a rising but still weaker power.[44] His conclusion is therefore a pessimistic one, since neither China nor the United States are likely to be capable of limiting their military space activities given their overarching struggle for hegemony.

Others see that China's current, largely accommodative policies toward the United States as a strategy of "buying time" rather than any real shift in Chinese attitudes toward its ultimate goal of replacing the United States as the regional (if not global) hegemon. Some analysts make the case that Chinese leaders may currently believe that these policies are indeed the best ones but that Beijing may *alter* or abandon restraint when the time for a possible power shift comes. As Avery Goldstein argues, "those preferences may change as China's capabilities grow, either because power sparks greater ambition or because others react in ways that provoke China to reconsider its national interests."[45]

By contrast, the historian Bruce Cumings sees U.S. hegemony itself and not problems of Asia's own making as largely responsible for the region's current angst and sense of disenfranchisement. He describes Washington over the past several decades as a calculating and exclusionary power in Asia that has fomented or exacerbated many of the region's longstanding sources of instability to its own benefit.[46] Specifically, Cumings blames the

United States for failing to settle such problems as the status of Taiwan and the division of Korea, both legacies of the cold war. He sees Washington as seeking to promote regional tensions as a dominant hegemon, arguing that "East Asian leaders are still on the outside looking in."[47] Given the U.S. failure over the past decade to propose new formulas for space security, according to this argument, the U.S. preference for bilateral, hub-and-spoke relations may be exacerbating regional space tensions.

Traditional Realism and Balance-of-Power Theory

Time-proven factors of mistrust, nationalism, and military competition are seen by other analysts as obstacles to regional cooperation. Besides nuclear and missile programs possessed by a number of countries in the region (including China, North Korea, India, and Pakistan), space weapons may play a divisive role in rival "balancing" (via increasing military expenditures and/or alliances) or instead "bandwagoning" with existing or rising hegemons.

The Indian analyst K. K. Nair argues that China's military space program has "out-distanced every other Asian space power to the extent that it has decisively altered the 'balance of power' overwhelmingly in its favor and is likely to tilt the scales further in the next few years."[48] He concludes, in regard to regional security, that "an 'action-reaction' effect reminiscent of the nuclear order" could result in Asia, turning Asia's current civilian space race into an actual arms race.[49]

Such dynamics continue to influence Sino-Indian relations, despite recent efforts by the two sides toward at least a limited rapprochement. China's space activity is seen by Indians as threatening from the perspective of regional prestige and because of its recent security implications. Dipankar Banerjee, for example, views Chinese space plans ominously, with the 2007 Chinese ASAT test shifting Asia's space competition into a military-led direction.[50] Given China's peaceful statements about space up until that point, Beijing's traditional support for strengthened arms-control measures, and its apparent focus on human spaceflight activities, Banerjee calls the test "both a surprise and a shock."[51] Looking ahead in anticipation of reactions by Asian militaries, he warns, "the possibility of a destabilizing arms race in space could well be a likely fallout."[52] India's announcement in 2010 of its plans to deploy a weapon capable of destroying satellites in low-Earth orbit suggests that traditional action-reaction military dynamics have already begun to manifest themselves among Asia's space powers.[53]

While the main focus of Asian security watchers is on China and India, some analysts argue that Japan's changes should not be ignored.[54] Indeed, while most analysts have assumed that Japan will serve as a calming influence in any future conflict over space, changes in Japanese society and a more assertive nationalism and militarism suggest that Tokyo is no longer simply a status quo power. In discussing the postwar (Heisei) generation of Japanese and their attitudes, Kenneth Pyle argues: "The new Heisei generation of politicians . . . believes that Japan must assert its own identity in international society."[55] Regarding China, he concludes that "The Heisei generation sees China not as a war victim but as a rival."[56] If such ideas take hold in matters of policy, disputes in areas such as territorial matters, the environment, and space may become confrontational.

Others point to both historical animosities and conflicting twenty-first-century strategic interests to explain the failure of regional institutions in Asia (compared to Europe), particularly in the security arena. Gilbert Rozman, for example, uses the term "stunted regionalism" to describe relations among the countries of Northeast Asia,[57] arguing that this subregion has not fulfilled its potential because of a number of enduring historical, military, economic, political, and strategic differences among the major countries. As Rozman observes, "The Chinese are framing a continental strategy appealing to South Korea to join. In contrast, the Japanese are distancing their country from it, seeking to counter with a maritime strategy reliant on the United States, and calling on South Korea to complete the alliance structure from the Cold War."[58] These fissiparous tendencies present obstacles to the future formation of effective regional organizations.

Nevertheless, Michael Swaine argues that "no concrete, interest-based set of national imperatives . . . exists to drive Beijing and Washington toward strategic rivalry over at least the medium term."[59] Yet Swaine is still not optimistic about emerging trends, suggesting possible conflicts on the horizon. He notes: "Washington should continue to discourage, or in some cases prevent, Beijing's acquisition of military capabilities . . . that directly challenge U.S. military superiority."[60] Although Washington did not try to dissuade China from conducting its harmful ASAT test in 2007, it has since tried to engage Beijing in space security discussions over this and other matters.

Economic and Neoliberal Approaches

The argument that cooperative organizations can smooth international relations by codifying acceptable behavior, providing a legally based forum for the resolution of disputes, and reducing resort to military force has become increasingly accepted since the late twentieth century. This has occurred despite the relative decline of U.S. hegemony, which was seen as the founding force for many post–World War II international institutions.[61] The increasing scale of international trade and investment has helped to drive this process, in part due to the self-interest of states. These trends have become even further engrained since the rapid expansion of international telecommunications in the 1990s, following the creation of the Internet, and associated changes in banking and finance that have globalized the investment community. Leading members of this school of thought have posited that military conflict may become increasingly outmoded as borders blur and the world becomes "flat" in terms of the removal of obstacles to economic interactions and thus international interdependence.[62] Indeed, the act of joining international institutions, such as the World Trade Organization (which links China and the United States) involves a compromise. As Robert Keohane summarizes, "Committing oneself to an international regime implies a decision to restrict one's own pursuit of advantage of specific issues in the future."[63]

Vinod Aggarwal and Min Gyo Koo argue that in the twenty-first century, Asia's leading powers have begun to move increasingly toward an embrace of trade liberalization as a collective (or club) good for pursuing mutual economic benefits.[64] Moreover, they argue that the increasingly creaky foundations of the San Francisco system in security affairs (based on the post–World War II treaty) has "prompted Northeast Asian countries to deal with new security challenges by collectively pursuing security cooperation as club goods," such as the Six-Party Talks concerning North Korea.[65] Drawing on the positive experience of free-trade associations (FTAs) and in light of the increasingly outmoded U.S. hub-and-spoke security framework in Asia (given the extent of transregional problems),

Aggarwal and Koo see a trend in which "countries are now pursuing greater institutionalization at the regional level, actively weaving a web of FTAs and security dialogues."[66] If this is true, a key question facing neoliberal institutionalists is whether such security cooperation is sustainable in key evolving fields such as space, where technology is changing at a rapid clip and there is little current evidence of dialogue.

In fact, a number of authors fault Washington for failing to work hard enough toward conflict prevention and institution building in Asia, both in regard to space and more generally in the security field, and preferring instead its traditional bilateral ties with a number of regional partners and allies. As Jonathan Pollack warns, there could be costs to this inattention:

> In the event of an insufficiently attentive US regional policy, leaders across Asia and the Pacific could begin to view the United States as a more distant power that focuses on the region only when vital US interests are at risk. Under such circumstances, the United States may find itself progressively less attuned to the regional future, and hence less able to shape events to its advantage.[67]

The same argument may hold for space. Fortunately, there is less "existential" animosity than existed during the cold war—in terms of fears of systemic domination or defeat. But as Muthiah Alagappa observes: "Power and force still play a key role in the interaction of the major powers, although their function is primarily in the defense, deterrence, and reassurance roles, not in conquest, domination, and territorial annexation."[68] Alagappa goes on to say that while Asian progress has been made in terms of joining international arms control and nonproliferation regimes, "Nevertheless, the overall record is quite poor, and the same story goes for transparency and accountability in matters of defense."[69] He summarizes, regarding Asian military relations, that "There are few bilateral arms control agreements and virtually no significant regional arms control accords."[70] These conditions hold for space as well.

Johnson-Freese sees the main emerging challenges in space from the perspective of economic globalization and the objective demands for increasing cooperation. As she argues, "Fundamentally, globalization is all about connectivity."[71] She cautions that highly restrictive U.S. export controls in regard to space and a unilateralist attitude toward space security during the Bush administration (2001–2009) have created a situation in which "countries are increasingly and deliberately choosing not to work with the United States" in space.[72] As a remedy, she proposes that the "United States must acknowledge other nations exploring and utilizing space as potential collaborators and view partnerships with them as opportunities to exhibit our leadership in the global commons."[73]

In regard to the views of other states in the region, David Kang argues that there is an "absence of balancing against China," given the tendency among at least East Asian states "to share a view of China that is more benign than conventional international relations theories might predict."[74] Indeed, China has been calling for arms control in space at various international fora, including the United Nations' First Committee and at the Geneva-based CD. Yet—as will be discussed in chapter 3—China's reluctance to give up its own ASAT system has failed to win it the trust of core Asian countries in regard to space security, including particularly India and Japan.

Constructivist and Ideational Approaches

Rather than focusing on national interests, power-based principles, or concepts of economic self-interest, a final group of analysts has focused on the role of ideals and social values in explaining Asian security policies and perspectives. In this literature, authors have emphasized self-identification and socially constructed beliefs as key determinants of national behavior.

In describing their vision of future security decisions in Asia, for example, Chung-in Moon and Seung-Won Suh admit some role for power balancing as a factor in the Asian security scene[75] but argue: "More critical is national identity and the politics of nationalism, which are closely intertwined with collective memory of history. National identity is important because it is responsible for shaping shared norms, interests, and images of countries in the region, eventually affecting mutual policy behavior."[76] Unfortunately, this emphasis leads them to take a relatively pessimistic view of prospects for near-term cooperation. They write: "The unresolved clash of identity politics has reinforced the security dilemma," making the peaceful settlement of differences at best uncertain.[77] In this regard, Minxin Pei points out a related problem in seeking U.S.-Chinese rapprochement. Put simply, she says, the two sides have "profoundly different world outlooks, especially regarding the principles of sovereignty and international order."[78] This could well get in the way of any broader security settlement, particularly regarding space. In Veblen's terms, we could call this gap a missing part of the "culture" of the cold war's space race.[79]

Susan Shirk, however, cautions against reading too much intention and aggressiveness into China's military programs: "Although China may look like a powerhouse from the outside, to its leaders it looks fragile, poor, and overwhelmed by internal problems."[80] Yet she also warns that its very instability "should worry us," if these problems lead its leaders to resort to hard-line nationalism as a means of domestic governance. Instead, Shirk recommends U.S. positive engagement to "lavish respect on China's leaders," reassure their sense of identity, and "thereby diminish . . . the need for them to make themselves popular by whipping up assertive nationalism."[81]

A related perspective on Asia's security future that may bear on the stability of future space relations is the so-called democratic-peace theory. This approach argues that the gradual adoption of democratic-capitalist ideas and practices within the region will eventually transform it and create conditions where conflict is avoided *or* channeled into international institutions where they can be handled through peaceful, administrative mechanisms. Writing in 1989, Francis Fukuyama spoke of the post–cold war system, which was then just emerging, as a "de-ideologized world" and predicted that conflict and struggle "will be replaced by economic calculation, the endless solving of technical problems, environmental concerns, and the satisfaction of sophisticated consumer demands."[82] Solving space's challenges—in a world of ideological consensus—would seem to fall neatly into his three sets of activities. In regard to China, Fukuyama saw a positive future, noting "anyone familiar with the outlook and behavior of the new technocratic elite now governing China knows that Marxism and ideological principle have become virtually irrelevant as guides to policy, and that bourgeois consumerism has a real meaning in that country for the first time."[83] Yet China's continued failure to adopt democracy, despite its embrace of market capitalism, has put the reliability of such predictions in some doubt. In other words, ideational consensus has not yet emerged and does not seem to be on the horizon.

Indeed, unexpected forms of competition even among those countries that have adopted democracy since World War II present troubling contradictions to the predictions of democratic-peace theory. As Paul Midford observes in this regard: "The lack of substantial security cooperation between Japan and [South] Korea and even more so the *de facto* alignment of Korean foreign policy with China against Japan on a number of issues poses an anomaly for the generalization that democracies align with each other."[84] Midford emphasizes the enduring role of history in Asian identity politics.

For some analysts, attempts to characterize the region through any purported ideational consensus are a fool's errand. As Stephan Haggard argues, the Asian region is instead "politically heterogeneous" and "it is far from clear what applicability the theory of democratic peace might have in such a setting."[85] Even if democratic factors were to become increasingly important, he observes, they might lead to heightened nationalism rather than greater regional cooperation.[86]

Thus, the ideational challenges facing the emergence of greater regional cooperation and institution building remain large. A key question addressed in later chapters will be what role space technologies and related political and strategic interactions might play in worsening *or* reducing these challenges to Asian regional governance and conflict prevention.

CONCLUSION

As neoliberal analysts have stressed, there is a growing appreciation within Asia of the necessity of continued engagement in the global economy, which may temper unilateral military responses to risk and foster collaborative efforts, particularly given market-based pressures for space industrial collaboration. At the same time, however, power-based theorists posit that meaningful cooperation in space will be difficult to accomplish, given space's links to both regional prestige and national security. The emerging role of military space activities within Asia also highlights the future risks of failing to address the current competition *before* space conflicts emerge and spill over into the military realm.

As noted above, this situation stands in contrast to the U.S.-Soviet space security relationship, which involved close communication—at least from the early 1960s onward—and a series of formal and informal restraint agreements. In this regard, the shift from bipolar dynamics to multipolar dynamics in space competition may risk destabilizing the space environment. Within Asia, an increasing number of spacefaring nations are vying for advantage in the absence of strong consensual rules of behavior or regular and substantive communication about space security. For this reason, today's Asian space competition requires additional analysis of national programs and perspectives. It is also important to understand why the current state of regional affairs has developed, what its characteristics are, and what might be done about it. The concluding chapter of this book, therefore, discusses policy options, with the aim of helping Asian countries and other spacefaring nations identify possible pathways out of their current dilemma.

CHAPTER TWO

THE JAPANESE SPACE PROGRAM

Moving Toward "Normalcy"

By any number of measures, Japan has long been the most accomplished space power in Asia. It has more than five decades of achievements in space science, has extensive practice in human spaceflight, and produces sophisticated launchers, satellites, and robotic devices equal to the world's best. However, Japan's space mandate formally excluded military space activities until 2008, putting Tokyo well behind leading powers in military applications and military operational experience. In addition, in the civil space area, while Japan's transfer vehicles now have independent access to the *International Space Station* (*ISS*), Japan has not yet made a decision to develop its own astronaut-rated space launcher. Finally, its commercial sector has only recently secured its first contracts for satellites and space-launch services, in part given the high cost of Japanese space technology.

Part of the reason for these limitations relates to its close ties to the United States, which occupied the country following Japan's defeat in World War II, shaped its peace-oriented constitution, and provided it early on with privileged access to space services and technology. U.S. military space programs have long shared information with the Japanese government, and the U.S. space shuttle has served as a transport vehicle for Japan's astronauts as well as its large hardware contributions to the *ISS* (including the *Kibo* module). But new security concerns—in particular North Korea's nuclear and missile programs—have kicked off a wide-ranging domestic debate on the proper role of space in national security and its broader contribution to Japan's power, prestige, and place within Asia. As a result, a Diet vote removed Japan's prior legislative restrictions against military uses of space. Japan's space program is now edging toward a more "normal" role, although the parameters of its activities in space are still being shaped by the country's unique perspective, priorities, and capabilities.

Given these trends, Japan's space program is truly in transition. It is moving beyond its prior cold war boundaries and into a more active and more international presence. Japan is well placed technologically to compete in space, but it has struggled to identify an effective organizational structure to govern its space activities and must deal with a budgetary framework in which its traditional support is only .25 percent of the national budget, far below that of the United States or China. It seems that despite its expanded focus on space, Japan is likely to struggle with the costs necessary to operate a much larger space program—one replete with independent human spaceflight capacity, a range of commercial services, and at least passive military assets. It has the potential, but it will have to weigh the role of space in its overall set of national priorities. These include budgetary limitations, legacy issues related to its pacifist constitution, and questions about the need to achieve full autonomy given the significant benefits available from its close cooperation with the United States. Nevertheless, Japan's unstated rivalry with and concerns about China's rising space capabilities—as well as missile and WMD threats from North Korea—are challenging Japanese decision makers to move in directions that previously seemed unapproachable. Japan's private sector is also becoming more forceful in promoting its space interests.

Japan's motivations include both civilian and security-related components. According to the Japanese political economy expert Saadia Pekkanen, Tokyo has long targeted space for special attention, because of the view within government organizations related to space "that this leading industry of the twenty-first century will spur transformations in existing technologies, underwrite the future of the information society, and expand the frontiers of humankind."[1] At the same time, Japan's younger generations have become increasingly committed to normalizing Japan's role in security affairs and achieving greater independence in ensuring its own defense within the Asian region. The Japan scholar Kenneth Pyle observes a more self-confident approach among Japan's emerging elite, evidenced by the recent upgrading to ministry status of the Self-Defense Forces and a more general belief that "Japan must assert its own identity in international society."[2] For some, this means constitutional change to allow collective self-defense and even the use of force overseas; for others, it means changing existing legislative restrictions to take a more active role only in UN-sanctioned activities. Space plays a part in this equation because of its critical informational and force-support role. In addition, Japan's acceptance of a military space program may reflect bottom-up pressure from industry. As Pekkanen and Paul Kallender-Umezu argue: "A specific set of Japanese corporations has shifted more visibly than ever before from commercialization to militarization of space-related technologies—a process that has had and will continue to have important implications for Japan's space military capabilities for national defense."[3] Such trends raise serious questions about the reactions of Japan's Asian neighbors, who remain highly suspicious of Tokyo's motives, given the weight of twentieth-century history.

This chapter discusses the evolution of Japan's space program, from its roots during World War II to the modern, multibillion dollar industry it is today. It explains the gradual expansion of the space program's charter from a narrow scientific effort, to a larger commercial endeavor, and finally to a more fully fledged national security effort with an explicit military component. In doing so, this chapter analyzes the domestic foundations for these changing space policies and their international implications. As Hirotaka Watanabe argues, "autonomy and international cooperation have been the central pillars of Japan's space policy," even though these goals have "often contradicted each other" historically.[4] Japan's active cooperation with the United States has been both a benefit and a crutch over the past several decades. Tokyo decision makers are now facing the challenge of whether to assert greater autonomy and regional clout—via a more active and possibly independent human spaceflight program and more robust military capabilities—versus the costs, risks of failure, and possible exacerbation of regional tensions. Japan has ambitions to be at least a regional leader in space exploration, as evidenced by its establishment and sponsorship of the Asia-Pacific Regional Space Agency Forum. But China's rise in space and Japan's post-1945 failure fully to convince its neighbors of its benign intentions have created challenges for Japan in asserting a broader space presence. For these reasons, the chapter makes the case that

Japan's space policy is currently conflicted. It would prefer to follow a cautious, step-by-step approach as it transitions to a more active role in orbit, balancing its desire for new military capabilities with its need to maintain good relations with the rest of Asia. Yet its changing politics and uncertain factors in its security relations with China make a stronger defense-oriented approach possible as well, with likely regional implications.

THE ROAD TO SPACE: FROM KAMIKAZES TO PENCIL ROCKETS

Just as in a number of countries in the 1930s, Japan had rocket enthusiasts and people who dreamed of spaceflight. Some met in a small organization known as the Nippon Rocket Society.[5] There is little evidence, however, of active experimentation with hardware outside of some limited efforts within the Japanese military and nothing on the scale of the German, Russian, or American programs. Thus, the real roots of the Japanese space program can be traced to World War II when—like a number of other combatants—Japan began to experiment with the potential use of rockets for assisting the takeoff and flight of aircraft, for supplementing artillery, and for possible use in longer-range attacks against ships, cities, and military units.[6] The man who would emerge later as the leader of Japan's rocket program, Hideo Itokawa, spent the war working for the Imperial Army as an engineer developing fighter aircraft at the Nakajima Aircraft Company, after graduating in aeronautics from Tokyo University.

But Japan lagged behind the world's leading rocket powers in terms of technology. It focused its main efforts on solid-fuel propellants, which proved difficult to scale up into anything approaching Germany's V-2 missile, used extensively in the European theater of war. Japan developed only a few short-range systems for barrage attacks against enemy troops. As the war wore on, however, the Japanese sought to reverse their fading chances of victory against the United States and Britain in the Pacific theater by using rockets to increase the effectiveness of kamikaze attacks against allied surface ships. In the summer of 1944, the high command acted on an idea provided by a young navy ensign and began designing prototypes for a high-speed, solid-fuel, rocket-powered kamikaze aircraft called the Ohka (or "cherry blossom") under the cover of the secret Marudai Project.[7] The Ohka would be brought to the battlespace lodged within the open bomb bay of a modified Mitsubishi G4M aircraft. Then, when within range of enemy ships, it would be released and glide until its pilot could establish his bearings and ignite the Ohka's onboard rocket motor, sending it hurtling toward its intended victim.[8] It would avoid antiaircraft fire by traveling at speeds nearly double that of normal planes. The pilot would detonate the Ohka's large, high-explosive payload moments before colliding with the target vessel.[9] These deadly devices, designed by Tokyo Imperial University's Aeronautical Research Institute, were used against a number of U.S. ships involved in the battle for Okinawa in April and May 1945, causing significant damage and casualties.[10]

A second, liquid-fuel rocket technology also entered into the Japanese military program in 1944, although this one by technology transfer. With the fate of the Axis powers becoming increasingly precarious, Nazi Germany sought to help its wartime ally by passing technical designs for the rocket-powered Messerschmitt 163 fighter plane via a long-distance Japanese submarine mission. These designs reached the Mitsubishi Corporation in mid-1944 and helped Japan develop a prototype aircraft by the summer of 1945, using a volatile liquid-fuel motor. But tests of the aircraft resulted in a series of problems and explosions, and the experimental fighter never saw service.[11]

After Japan's defeat in August 1945, the allied powers banned the construction of both aircraft and rockets by Japan.[12] The engineers who had worked on these programs turned to peacetime employment. But Japan's adoption of its new, U.S.-authored constitution rejecting the use of force in 1947 and its agreement to the terms of the San Francisco Peace Treaty in 1951 restored Japanese sovereignty and led to the removal of the postwar industrial restrictions on rocketry.[13]

The University of Tokyo's Institute of Industrial Science resumed its place at the center of rocket research, under a new organization called the Avionics and Supersonic Aerodynamics Research Group. Professor Hideo Itokawa, who had returned to rocket research after spending the early postwar period studying acoustics, became head of the new group.[14] In their resource-strapped conditions, the faculty and students focused on inexpensive solid-fuel designs for their early rockets.[15] With government support from the Ministry of Education, they began extensive testing of the so-called Pencil rockets, whose initial prototypes measured less than ten inches tall. By 1955, the institute's team had relocated its launch site to

Michikawa Beach in Akita Prefecture for launches of the larger, so-called Baby rockets, reaching altitudes of up to six kilometers.[16]

JAPAN'S FIRST SATELLITE

With rocket scientists internationally gearing up for the International Geophysical Year (IGY)—a worldwide technical exchange scheduled for 1957–1958 modeled on shared efforts to further polar research earlier in the century—Japanese rocket specialists sought state funding for a sounding rocket program. The government finally offered its support for the Kappa sounding rocket (named after the Greek letter) in the wake of the Soviet Union's successful *Sputnik* orbital flight in October 1957.[17] The team's Kappa-6 rocket flights became Japan's contribution to the IGY, collecting information from thirteen separate launches to altitudes nearing sixty kilometers about cosmic rays and conditions in the upper atmosphere.[18] A series of more powerful Kappa rockets followed, including the Kappa-8-1 in July 1960, which entered space (at one hundred kilometers), reached an apogee of 190 kilometers, and conducted the world's first measurements of ion density outside the atmosphere.[19] Organizationally, the accomplishments of the rocket scientists during the IGY led the Japanese government to create a National Space Activities Council in 1960 under the prime minister.[20] In the same year, Japan held its first International Symposium on Space Technology and Science in Tokyo, seeking to publicize Japanese developments and promote research exchanges with leading foreign space scientists. By the end of the 1960s, twenty countries would send representatives to this increasingly well-attended international meeting.[21]

In order to take the next important steps in space exploration, the Japanese government supported the creation of the specialized Institute of Space and Aeronautical Science (ISAS) at the University of Tokyo in 1964, which became the new national organization for space research. Japan had now completed its Kappa series and moved to the more powerful Lambda rockets, whose L-3 version had by now achieved altitudes of one thousand kilometers.[22] ISAS aimed for still greater thrust and possible orbital insertion with the follow-up Mu-series launchers. For this effort, the Japanese government established a new launch site at Kagoshima on the island of Kyushu in the south of Japan in 1962. The launch site's eastward flight direction over water reduced the risk to Japan's population. But ISAS personnel faced protests from fishermen for scaring away their prey and from local residents for using radars that interrupted their radio programs.[23] Nevertheless, ISAS continued to make progress. But the challenges of spaceflight proved daunting: the Mu rocket was delayed, and from 1966 to 1969, ISAS failed four consecutive times in efforts to launch an orbital satellite with advanced, four-stage Lambda rockets.[24] These problems eventually forced Itokawa to resign from ISAS.[25]

NASDA AND THE INTERNATIONALIZATION OF THE SPACE PROGRAM

During the 1960s, Japan began to reach out to various international bodies in order to raise its standing, normalize its postwar status, and begin to integrate its space scientists into the broader space science community. It joined the U.N. Committee on the Peaceful Uses of Outer Space in 1962 and later became a member of the International Council of Scientific Unions' Committee on Space Research,[26] to facilitate data exchanges and Japanese participation in international meetings regarding space.

Domestically, the government decided to create two new organizations for space activity: it restructured and elevated a renamed Space Activities Commission within the prime minister's office in 1968, as the senior advisory body responsible for coordinating space policy, and created the National Space Development Agency (NASDA) in 1969.[27] ISAS still remained in charge of small scientific satellites and solid-fuel sounding rockets but began to fade from public and governmental budgetary attention. As Harvey notes, "Japan in effect developed two different, parallel space programmes, a unique situation."[28] NASDA was charged with implementing civil space activities, including satellite development, major launcher development, and the operation of launch facilities and tracking and control stations.[29] In addition, it provided a public face for interacting with foreign space agencies and companies, which had become important for Japan in order to receive technology for liquid-fuel rockets being offered by the United States. Fearing that this new rocket technology might be diverted for ballistic-missile purposes,[30] however, the Diet stepped in with a resolution in 1969 that required Japan's space program to be conducted exclusively for nonmilitary purposes and through nonmilitary means. Agreements signed in 1969 with the United States also forbade Japan from transferring launch technology to third parties.[31]

In February 1970, Japan became just the fourth country (after the Soviet Union, the United States, and France) to orbit a satellite around the Earth, with its small *Ohsumi* spacecraft aboard a solid-fuel Lambda 4S-5 rocket.[32] With this flight, Japan became the first Asian nation to join the space age, doing so with a purely scientific and civilian rocket developed using its own technology. In 1972, Japan used its newfound status to reach its first international agreement beyond the United States when Tokyo began regular meetings and a set of cooperative projects with the European Space Agency.[33]

In order to expand its capabilities, the next steps for Japan's space program were the mastery of more controllable (and more complex) liquid-fuel rockets and the development of satellite communications. For these tasks, Japan looked to its ally, the United States, for technical assistance under its existing agreement. NASDA began work on a new rocket, the three-stage N-1, its first major liquid-fuel rocket. While Mitsubishi Heavy Industries served as the main contractor, the N-1 was based on the U.S. Delta (Thor-derived) launcher developed for NASA via licensed production arrangements with the airframe manufacturer McDonnell-Douglas, the engine makers Rockwell and Thiokol, and the control-system designers TRW and Honeywell.[34] For some Japanese involved in space activities, the deal represented a significant opportunity for Japan to acquire valuable new technology. But others in the space program viewed the project as "a political ploy [by the United States] to pressure Japan to increase its dependence on US technology."[35] As the Japanese space analyst Kazuto Suzuki explains, "The people at ISAS considered it a deliberate attempt to destroy Japanese endogenous technology."[36]

In a further break with ISAS, NASDA now centered its flight operations at a new launch site located even further south, on Tanegashima Island. But, in accommodating Japan's unique socioeconomic culture, NASDA still had to limit its launches so as not to disrupt the local fishing industry.[37] Japan's space budget benefitted from what two NASDA officials described as a "remarkable increase" in the early 1970s, reaching $250 million by 1975, as the government made a strategic decision to push forward toward making Japan a significant space power.[38]

The Japanese government also teamed with a variety of U.S. companies to purchase communications satellites, which were launched by the United States. NASDA succeeded in launching its first N-1 rocket in September 1975, placing the small and domestically built *Engineering Test Satellite*, or *Kiku-1*, into low-Earth orbit.[39] NASA assisted Japan in providing tracking data. After additional work, NASDA succeeded in placing *Kiku-2* into geostationary orbit 22,300 miles over Japan, becoming the third country in the world to accomplish this task.[40] This launch set the stage for the subsequent indigenization of Japan's telecommunications satellite industry and the establishment of independent satellite-assisted meteorological and Earth-observation capabilities.

In 1978, the Space Activities Commission issued an ambitious fifteen-year plan entitled the "Outline of Japan's Space Development Policy." The guidelines even envisioned an independent human spaceflight program that would carry out such industrial tasks as space-based materials processing in zero gravity, but a 1984 revision linked Japan's manned activities instead to the planned U.S.-led space station.[41] The main focus of the 1978 plan—the establishment of national technical capabilities to carry out "comprehensive space programs"—continued. Part of this effort included plans for a heavier launch booster, the N-II, which made its first successful flight in 1981, and an even more powerful H-I, approved for development that same year. Although 84 percent of the vehicle would be designed and constructed in Japan under an overall contract won by Mitsubishi (with participation by Ishikawajima-Harima Heavy Industries and Nissan Motors),[42] the H-I would be another U.S.-derived vehicle, this time using more advanced cryogenic technology and liquid-hydrogen fuel.[43] Licensing arrangements with the United States, however, would prevent Japan from marketing the H-I commercially.[44]

ACHIEVING A MATURE CIVIL SPACE PROGRAM

By the mid-1980s, Japan had established its own infrastructure for constructing and launching communications and broadcasting satellites, and images from its meteorological satellite (*Himawari*) had become "quite familiar to the public of Japan through the daily TV weather programs."[45] Although still divided between ISAS and NASDA, both programs continued to make progress. ISAS's establishment of the Usuda Deep Space Center in 1984 opened up new capabilities for communicating with distant spacecraft and facilitated Japan's high-profile, dual-satellite mission to study Halley's Comet in 1985 and 1986.[46] On the civil side, NASDA flights with the H-I rocket began in 1986. The series was marked with considerable success, launching a total of nine satellites in flights through 1992.[47] However, Japan wanted to move on to a new domestic design, the H-II, in order to achieve full technological independence and the ability to pursue the international commercial market for launching satellites into geostationary orbit.

One of the major shifts for Japan during this period involved Japan's preparations for participation in the human-tended research projects associated with the U.S. space shuttle program. In April 1984, Japan began the selection process for its first astronauts, forming an initial group of sixty-four candidates (including five females) from an initial applicant pool of 533.[48] In 1985, Japan selected three payload specialists to serve as its representatives in space. Japan's formal participation in what would later become the *ISS* began in 1987 with the signing of the Intergovernmental Agreement with the United States. A major part of Tokyo's participation centered on its pledge to contribute the so-called *Japanese Experimental Module* (later known as *Kibo*) to serve as a test facility for materials processing in space. However, the official space program's efforts to put the first Japanese into orbit aboard NASA's space shuttle were bypassed in December 1990, when the Tokyo Broadcasting System paid for a flight by the forty-eight-year-old journalist Toyohiro Akiyama aboard the Russian *Mir* space station.[49] With NASDA astronauts already in training for upcoming U.S. shuttle flights, the *Mir* interlude was an embarrassment. But NASDA soldiered on, resolutely refusing to recognize Akiyama's accomplishment.[50]

The challenges for Japan in moving from its previously limited status into the wider commercial space arena and human spaceflight did not meet with universal domestic approval. One science writer observed in 1990: "Unable to define its goals clearly, the space program has largely failed to fire the imagination of the technologically literate Japanese public."[51] Noting the flat space budget of about $900 million for the past nine years, he concluded, "Japan is entering its third decade of space activities more confused than ever about where to proceed next."[52]

To support its space station plans, NASDA began an additional astronaut selection process in 1991 to create the corps for future flights to the eventual orbital station and NASDA's own planned module. Finally, in September 1992, the NASDA astronaut Mamoru Mohri, a Ph.D.-trained chemist, became Japan's second man in space by flying aboard the U.S. space shuttle *Endeavor*, where he conducted a series of life-sciences experiments. Mohri's flight began a series of five additional U.S. shuttle flights by NASDA astronauts by the end of the 1990s, including the first Japanese woman, Chiaki Mukai (a doctor with an additional Ph.D. in physiology), who flew in July 1994 and again in October 1998.[53] After the perceived success of the U.S. space shuttle and the Soviet *Buran*, NASDA and Mitsubishi issued a report in 1994 calling for a reusable Japanese space plane, which gained government research support under the planned name *Hope* in 1995.

Meanwhile, ISAS continued its unmanned sounding rocket program in parallel to NASDA's work in civil space. In 1990, Japan began a series of lunar missions operated by ISAS, becoming the third country to place a satellite into lunar orbit, despite what one space writer at the time called a "shoestring budget of $41 million."[54] As ISAS Director Jun Nishimura explained, "Our philosophy is that good space science does not have to be expensive space science."[55] Although these early missions were not entirely successful, due to communications and control problems, Japan also became the third country in the world to send an object to the Moon's surface, when it directed its *Hiten* satellite into a crash landing in April 1993.[56] By 1996, ISAS had also launched 325 sounding rockets to conduct research in such areas as microgravity, ozone depletion, and energy transmission.[57] ISAS's research on the polar ionosphere and auroral phenomena included launches from Norway and dozens from Antarctica (beginning in 1971).[58] Despite its successes, the relative share of ISAS's budget in the total Japanese space budget continued to decline as the expenses for NASDA's human spaceflight activities and extensive satellite operations grew at a more rapid pace.

At the same time, Japan also began to take a more proactive political role regarding space cooperation, spurred in part by a resolution passed at the Asia-Pacific International Space Year Conference in 1992, which called upon states to promote regional space cooperation and data exchanges.[59] At a September 1993 meeting co-hosted by NASDA and ISAS, Japan established the Asia-Pacific Regional Space Agency Forum (APRSAF), committing it-self to a series of yearly conferences and Japanese-led training programs and data exchanges aimed at promoting regional cooperation in space—as well as Japan's role as a leader. Among APRSAF's stated objectives were the use of space to "contribute to socio-economic development [in] the Asia-Pacific region and the preservation of the global environment," "exchange views, opinions and information on national space programs and space re-sources," and to "discuss possibilities of future cooperation amongst space technology developers and space technology users to bring mutual bene-fits [to] the countries in the Asia-Pacific region, [and] identify areas of common interest."[60] APRSAF held its first four meetings in Tokyo but be-gan to reach out further to the rest of Asia beginning in 1998. The organi-zation began to raise the profile of Japan's NASDA and eventually initiated a series of training efforts for less-developed Asian nations, assistance pro-grams (such as the provision of telescopes, satellite data, and ground sta-tions), and eventually joint development projects.

In the commercial realm, Japan's domestic space industry had benefit-ted since its inception from preferential treatment by the United States in terms of trade: gaining access to U.S. technology but not being forced to provide entry into the Japanese market in exchange. But general political and economic developments in the late 1980s brought growing U.S. pro-tests over Japan's burgeoning trade surplus and closed markets, which in-cluded the market for domestic communications satellites. As a result, in 1990, the Japanese government agreed to lift import barriers that had pro-tected its satellite producers under a joint memorandum with the United States on "non-R&D satellites."[61] Overnight, Japan's satellite firms faced withering competition from more advanced foreign (especially U.S.) aero-space companies, who soon dominated the country's satellite communi-cations market, rendering a significant blow to the Japanese aerospace industry.

Japan also faced new challenges in its attempts to complete the indige-nous H-II rocket. The complex engine requirements resulted in two major explosions during testing that delayed the program and its first launch until 1994.[62] But the flight succeeded, giving Japan a seemingly reliable (albeit expensive) new launcher capable of delivering spacecraft both to low-Earth and geostationary orbits.[63] Although its second flight failed to deliver its Japanese-made satellite into the proper orbit, four subsequent flights through 1997 succeeded in launching satellites, leading it to be recognized as "one of the most advanced rocket systems in the world."[64] Indeed, in 1996, Hughes posted an order for ten launches of a still-planned and less costly H-IIA rocket;[65] this was followed by a similar order from Space Systems/Loral.[66] A highly promising commercial future finally seemed to be materializing for Japan in space. Unfortunately, in the next decade, the Japanese space program would experience a number of growing pains, causing officials to reconsider their strategies and forcing the country to recommit itself to space activity. The timing proved particularly inopportune: these problems coincided with the 1997 Asian financial crisis and the bursting of the Tokyo real-estate bubble, which placed strong downward pressure on the Japanese space budget.

CRISIS AND REORGANIZATION

The first of the problems for Japan's space program came in regard to its highly touted H-II launcher. In February 1998, problems in the H-II's second-stage cryogenic engine led to a failure to place a communications satellite into geostationary orbit.[67] The H-II's next flight in November 1999 resulted in a first-stage failure and the loss of another satellite, damaging the H-II's reputation internationally.[68] NASDA reluctantly cancelled the H-II series, putting its hopes on the H-IIA, still in production. Hughes, however, decided it could not afford to take chances and, in 2000, scrapped its prior order, as did Space Systems/Loral. Since the commercial cost of the H-II was "twice that of an equivalent launch by *Atlas* or *Ariane*,"[69] NASDA sought to redesign a less expensive vehicle by using U.S.-supplied solid-rocket boosters, reducing weight, and simplifying the production processes. The H-IIA would not have its first successful flight until August 2001.

Another motivating factor for changes in Japan's space program was North Korea's launch of a Taepodong I rocket over Japan's home islands in August 1998. Japan's complete dependence on the United States for reconnaissance information led senior Japanese officials to push for a national Earth-observation capability, which could be used for military purposes but had to be designated as "multipurpose" and operated by civilians in order to comply with the 1969 Diet resolution.[70] As the senior Liberal

Democratic Party official Takeo Kawamura commented, there were other perceived benefits of starting what became the so-called Information Gathering Satellite (IGS) program: "National security was part of the motivation, and from a wider point of view in terms of securing new sources of energy, our agricultural policy, international cooperation, and disaster prevention and monitoring, we needed to have a national strategy for the role of space in dealing with these issues."[71] In fact, as early as 1991 the government had started feasibility studies and a 1993 defense advisory panel had called for development of an Earth-observation satellite.[72] But Japan had no experience in space-based intelligence. In the end, Japan awarded a contract to Mitsubishi Electric Corporation to construct the reconnaissance satellites but worked with U.S. firms to obtain subsystems. To operate the system, however, Japan had to create a new organization in 2003 under the Cabinet Secretariat to manage the project, whose imagery analysts requested and received training in the United States.[73] Japan succeeded in launching the first two satellites for the planned constellation in March 2003, a one-meter-resolution electro-optical satellite and another with one-to-two meters of resolution using synthetic-aperture radar.[74] But NASDA's attempted launch of two follow-up satellites to complete the initial constellation failed, destroying the satellites and raising the costs of the project's already large $2.3 billion budget.[75] The accident also represented a major political embarrassment for NASDA and the space program.

In the face of growing frustrations over many years with the divisions in the Japanese space program and the rash of recent technical and oversight problems that had begun in the mid-to-late 1990s, the administration of Prime Minister Ryutaro Hashimoto began to reconsider ISAS and NASDA's bifurcated structure. These reforms were part of a statewide push for "small and efficient government" in the late 1990s,[76] which also sought to address problems of corruption and bureaucratic unresponsiveness and reduce costs. For the space program, this would result in two major organizational changes over the next ten years. In the first stage, in 2001, the government merged the Ministry of Education (responsible for ISAS) and the Science and Technology Agency (responsible for NASDA) into a single "super" ministry, now called the Ministry of Education, Culture, Sports, Science, and Technology (or MEXT). In the second stage, in October 2003, Prime Minister Junichiro Koizumi merged ISAS and NASDA into the Japan Aerospace Exploration Agency (JAXA). Despite these changes,

continued problems of oversight after the November 2003 IGS launch failure, the perceived lack of a unified space plan, and growing worries about
Japan's security prompted the Diet—backed by high-level Liberal Democratic Party members, independent analysts, and industry representatives—
to begin a push for more fundamental reform. As Kawamura, who had
served as the first head of MEXT and headed the initiative in early 2005
after leaving office, explains: "This accident was the result of a failure of
responsibility and a lack of strategic planning. In the aftermath of the incident I came to realize that there was no organization in control and exercising leadership in space development in Japan."[77] After considering the
lack of focus in the program, the Kawamura initiative called for a shift in
emphasis from science to applications, a more streamlined administration, and the freedom to pursue military uses of space. In May 2008, the
Diet enacted the new legislation, allowing military uses of space for the
first time and refocusing the civil space program as well.[78] These changes
created a highly fluid situation in regard to space policy, shaking up past
practices and putting the overall direction and management of space activities into question.

Despite the perceived "space crisis" of the late 1990s and early years of
the 2000s, both ISAS and NASDA had projects already in the pipeline that
began to come to fruition by 2007, restoring Japan's space prestige and indeed moving it considerably forward. NASDA's ill-fated *Hope* spacecraft
had gone through a series of revisions by the late 1990s, as the government
sought to reduce its projected costs by creating a smaller, unmanned module.[79] The leading alternative design that emerged from this process became the so-called *H-II Transfer Vehicle* (or *HTV*). Its purpose was to ferry
up to six tons of supplies and equipment to the *ISS* and return trash and
other used items to burn up in the atmosphere,[80] thus building Japan's experience in spacecraft operation. It had the benefit of providing a possible
building block en route to a future human-rated vehicle. The *HTV*'s first
mission in September 2009 proved successful, providing a big boost to the
ISS project and to Japan's role as a participating member. With the planned
end of U.S. shuttle flights nearing, Japan's *HTV* now added to the only two
other existing cargo-transfer options for the *ISS*: the Russian *Progress* and
the newly tested European *Automated Transfer Vehicle* (*ATV*).

In the field of space science, ISAS's excellent track record was initially
marred by the failure of an ambitious ISAS mission to Mars. Its *Nozomi*

spacecraft, launched in 1998, had experienced directional and control problems, reaching Mars four years later than scheduled. The delay turned out not to be the most serious problem, however, as the spacecraft's arrival in December 2003 was soon followed by its failure to conduct the necessary maneuvers to allow it to enter Mars's orbit.[81] Thus, *Nozomi* passed the planet by and could not complete its intended experiments and observations. In May 2003, however, ISAS launched the first spacecraft—*Hayabusa* (or *Muses-C*)—scheduled to travel to an asteroid and return samples to Earth. The spacecraft succeeded in reaching the asteroid—an object discovered by a U.S. scientist in 1998 and appropriately named Itokawa (for the Japanese space pioneer) in 2000, touching down on its surface in November 2005. Although technical problems with a planned lander forced a delay in its departure and a significant revision of its mission plan, *Hayabusa* blasted off and began its return trip to Earth in April 2007, arriving in June 2010.[82] The jury-rigged mission succeeded in bringing some 1,500 tiny particles back from the asteroid for analysis in the first project of its kind.[83]

Eager to show an increasingly technically literate younger generation of Japanese the value of spaceflight, JAXA began an even higher-profile lunar orbital mission by launching the *Selene* (or *Kaguya*) spacecraft in September 2007.[84] The mission, whose goal was to map the surface of the Moon from an initial altitude of one hundred kilometers and transmit high-definition images for release via the Internet—proved to be a resounding success and made the grainy photos and jerky video segments from the U.S. Apollo era seem antiquated. *Kaguya*'s startlingly sharp images provided new definition to the Moon's surface and helped stimulate new attention to space among both Japan's younger generation and space enthusiasts around the world, boosting JAXA's reputation as one of the world's leading space agencies. The *Kaguya* orbiter finally ended its mission by descending and crash-landing on the Moon's surface in June 2009. Japan's *Greenhouse Gases Observing Satellite* (*GOSAT*), launched in 2009 to monitor atmospheric levels of carbon dioxide and methane, also showed that JAXA was addressing problems perceived as significant by Japan's younger generation.[85]

Finally, Japan's work on the *ISS* since 2008 marked another significant upgrading of its space presence, experience, and accomplishments. After a lengthy process of design and development, the first segment of Japan's

Kibo experiment module—built at its Tsukuba Space Center—was launched into space aboard the U.S. space shuttle, along with the Japanese astronaut Takao Doi, in March 2008.[86] The event proved to be a major success, with intensive Japanese media coverage. Two additional construction flights followed to complete the *Kibo* facility, which will serve as a primary experimental module for plant growth, materials processing, and other research, including projects aimed at space commercialization.

By 2009, JAXA had reestablished the Japanese space program's reputation and set itself on a higher plane of accomplishments. While it had not launched its own astronauts into space (as had China), the sophistication of Japan's scientific and research missions had impressed the world space community. Yet Japan's 2008 space legislation and the 2009 transition in the Japanese government to the leadership of the Democratic Party of Japan (DPJ) raised new questions as to the future of the space program and its direction, as well as to the sustainability of funding for its planned expansion into additional human spaceflights, commercial efforts, and military applications.

CURRENT SPACE POLITICS

Japan's space organizations have been on a rollercoaster since 2001. The Japanese Diet member Katsuyuki Kawai recounts that a study group of legislators concluded that Japan's space effort was overly focused on narrow scientific research and development and was not serving broader national interests.[87] The purpose of adopting the 2008 Basic Law, he observes, was not just security related. Instead, Kawai suggests five principles for Japan to follow in its future space policy: diplomacy, national security, industrial development, stimulating the national population, and contributing to human evolution.[88] Indeed, a primer on the new Basic Space Law distributed in 2008 by the Cabinet Secretariat's Space Headquarters only mentions national security as a minor item under the section "Improvement of Citizens' Lives, etc." in its list of the law's contents.[89] One recent projection for Japan's space activities outlines a proposed 25 percent increase in space spending, to a total of $26 billion spread over five years (2010–2014), with a significant planned boost for new activities under the Ministry of Defense.[90] Despite these optimistic projections, however, the yearly space budget over the past two years has in reality remained flat, because of ongoing financial concerns in Japan.[91]

A key question is how the DPJ will come down on the future status of JAXA and other key components of the space program, if indeed the party remains in power long enough to bring reform efforts to fruition. Currently, over ten ministries are involved in space activities in some way, and there is increasing pressure within the DPJ government to consolidate these various segments. As the U.S. space policy expert John Logsdon explains: "The idea is to get at least most of the space effort out from under the control of the R&D-oriented ministry, MEXT, and to focus that effort on projects driven by industry interests in selling space systems and services to global markets and by a desire to develop space capabilities relevant to Japanese security interests."[92]

One specific proposal for further centralization of Japan's space activities since the DPJ's election is to create a kind of "super" JAXA or so-called Bureau of Space, which would swallow up the activities of all other ministries, including perhaps those in the new Ministry of Defense. While organizationally elegant, such a structure would diverge significantly from practices around the world—such as in the United States, Russia, Europe, and India—where civil, commercial, and military functions are divided. Yet the former state minister for space development and now Foreign Minister Seiji Maehara explains that the goal of the proposed reforms is "aimed at increasing the transparency of Japan's space development strategy and policy, and to look at how to unify both our decision-making and budget under one authority."[93]

Whether the new space bureau will move forward or receive adequate staffing, authority, and resources remains uncertain. Since June 2008, Japan has had a minister for space within the cabinet, a position that has received increasing political attention. But the position still lacks a supporting ministry and budget authority, and so far it plays more of a planning and representative role for space than one with significant implementing power. Under the space minister, the Strategic Headquarters for Space is tasked with developing reform plans and, if adopted, implementing them. But with its relatively small staff and the limited availability of outside expertise on space from key Japanese bureaucracies, it remains unclear how such an effort would be carried out. If the past is any guide, the process of change is likely to be incremental rather than sudden, given the enduring

power of the underlying bureaucracies and the traditional Japanese preference for consensus decisions.

CIVIL AND COMMERCIAL SPACE DYNAMICS

According to the Japanese space expert Setsuko Aoki, the Basic Space Law
passed in 2008 undertook to change the focus of Japan's space policy from
scientific research to "user-oriented space applications."[94] She argues that
in the current environment, the "voice of industry" is much louder than
before.

After decades of government plans for space, Japan has in recent years
reached two important milestones in developing a successful commercial
space program. First, in December 2008, the Mitsubishi Electric Company
became the first Japanese company to win a commercial contract with a
foreign entity, when a joint venture between Taiwan's Chunghwa Telecom
and Singapore's SingTel placed an order for a telecommunications satellite.[95] Second, in January 2009 Mitsubishi Heavy Industries won Japan's
first commercial launch contract, when the South Korean government selected its H-IIA rocket to launch its *Kompsat-3* Earth-observation satellite.[96] In April 2007, the Japanese government had transferred the H-IIA
rocket to Mitsubishi Heavy to facilitate such sales[97] and to distribute more
responsibility to the private sector.[98] According to Aoki, these successes
helped stimulate the drive for the new Basic Space Law, as it was clear that
Japan would finally be moving beyond the confines of its national market
to providing international space services, and the government believed it
was necessary "to promote and assist the space industry as a national project."[99] Japan is now looking for external markets for ground stations and
remote-sensing services as well.

Despite its recent successes, the prospects for Japan's "industrialization"
of space are likely to remain limited. While the market for space services is
gradually expanding, international competition is fierce, and there are
many more providers than in the past. Japanese services are expensive
when compared to Indian, Chinese, and Russian options. However, reliability and political factors also come into play in such decisions, so Japan is
likely to have a small but slowly expanding commercial market. At the same
time, not all Japanese space products will be equally competitive. Japan's

long-scheduled Galaxy X (GX) rocket, touted initially as a "clean" alternative to traditional launchers because of its use of a liquid natural gas second stage, is being phased out in the face of low demand and high costs, which have doubled from original estimates.[100]

The Japanese government's focus on applications over science can be seen in its continued willingness to fund the so-called Quasi-Zenith Satellite System (QZSS), a GPS-augmentation plan for Northeast Asia. Like the European Galileo system, which backers hoped could be funded through private investment, Japan's QZSS has faced difficulties attracting the support necessary for its development and launch. As a result, JAXA covered the cost of launching the first satellite in October 2010 as a proof-of-concept experiment, hoping to attract further investment from potential private users.[101] The eventual plan is for three additional satellites in highly elliptical orbits that will provide continuous coverage over the region.[102] Japan hopes to attract eventual South Korean support for the project as well.[103]

An increasingly central program to Japan's space effort is its human spaceflight component. Part of the reason NASDA originally supported a Japanese place on the *ISS* was a belief that a manned presence in space represented "modernity" for the country and would open options for commercial advancement, particularly in the area of materials processing. Accordingly, the main emphasis of the *Kibo* module was on the identification and development of commercially useful techniques for agriculture, medicine, and industry. The Ministry of International Trade and Industry established a Space Industry Office in 1987 to help facilitate its participation in the *ISS*.[104] But foreign experience on the *Mir* and space shuttle has since dampened hopes of any significant breakthrough being accomplished on *Kibo*, although JAXA has not reduced its commitment to human spaceflight as a result. Indeed, Japanese officials cite polling data indicating solid public support for human spaceflight in the country.[105] JAXA and other government agencies view the need to stimulate younger people's interest in space as a national priority, given Japan's aging population and the rise of space competitors in Asia. A major issue of debate is whether to pursue an independent human space-launch capability. There is strong support in some quarters but also concerns about costs and risks. Japanese space officials are worried that the death of an astronaut could create serious negative implications for the future of the space program.[106]

Given the ongoing impact of recent financial difficulties and the growing costs of Japan's welfare state in the context of its aging population, Japan may not be able to afford large future increases in its space budget to cover such an expansion. Japan increased its space budget by 3 percent for 2011, but the impact of the earthquake and tsunami forced all government agencies to absorb a 5 percent reduction. Nevertheless, a strategic decision seems to have been made that continued growth in space capability remains critical to Japan's image, technological reputation, and security.

THE POSSIBLE DIRECTION OF MILITARY SPACE ACTIVITIES

Japan's efforts to develop its military space capabilities stem from at least three main security objectives: (1) to bolster its independent reconnaissance capability in regard to key threats, such as North Korea and China; (2) to facilitate communications with ships and troops deployed overseas on UN missions or mandates; and (3) to support its growing activities in the area of missile defense, which it officially joined with the United States in 2003.

But Japan faces a series of challenges as it seeks to organize itself for military space activities. One problem is that the country suffers from a serious lack of governmental and military expertise in the area, as a legacy of the 1969 Diet resolution. Put simply, military space was off limits for Japan, and very few officials have developed any significant knowledge of the subject, much less operational experience. This puts the government in a poor position to evaluate the ambitious plans of industry in regard to possible national military assets. As Watanabe observes, "the Ministry of Defense has just begun to study space activities since its National Institute for Defense Studies (NIDS) has only a few space specialists."[107] For these reasons, Watanabe and other experts recommend greater cooperation with the U.S. military on space activities, despite the more proindependence proclivities of the DPJ government.

A second and related challenge is trying to determine the realm of the possible for military space amid a welter of industry proposals. At a public forum on space in April 2009, a senior industry representative declared that Japan should commit itself to building and deploying space-based defenses, noting, "We have the technology, all that is lacking is the political will to do so."[108] Such talk would have been forbidden before 2008, but now it is voiced openly, despite the lack of any clear demand for such systems, any real plan for their development and operation within Japan's existing space structure, or any consideration of the likely regional implications of such a move. Yet government officials questioned about such ideas treat them as legitimate and worthy of consideration. This is despite their sharp break with long-established Japanese space policies, every other country's lack of such weapons, and their highly controversial status in the international debate over space security.[109] Since most senior Japanese space officials have come from economic-oriented ministries, they tend to have an insular viewpoint and are not aware of the international security context for such decisions. Instead, they comment that such systems would be acceptable as long as they were "defensive."[110] Another controversial proposal has been the possible deployment of an early-warning satellite to monitor for missile launches, although such a system would have to be deployed in geostationary orbit at considerable expense for it to have continual coverage over North Korea.[111]

The Japanese government has appropriated funds for the Ministry of Defense (MOD) to open a Space and Maritime Security Policy Office, although it has been forced to rely mainly on officers from the missile-defense area, given the absence of trained space personnel. The new office is tasked with studying and then charting "ways the MOD can utilize space, how to integrate satellites and related systems into the military's infrastructure, and planning of future military satellite systems for communications, imaging and weather monitoring."[112] Part of this effort will include the completion of Japan's constellation of reconnaissance satellites. The government has succeeded in launching two additional IGS satellites and another more sophisticated optical satellite, with sixty-centimeter resolution.[113] Notably, in violation of the UN Convention on the Registration of Space Objects, Japan failed to list the orbital parameters of any of these satellites, joining certain other military space powers that have skirted this international obligation in the past.[114]

Yet Japan has taken a strict compliance policy on other space treaties. After the January 2007 Chinese antisatellite test, for example, Japan was the only country to categorize Beijing's dangerous act as a "violation" of the Outer Space Treaty. Specifically, Japanese Prime Minister Shinzo Abe referred to Article IX of the treaty, which prohibits activities that could contaminate the space environment.[115]

But Japan appears to be cautious about pushing too hard toward a ban on debris releases in space, out of fear of constraining its own future missile-defense efforts[116] and because of pressure from the United States against a binding debris treaty (versus a voluntary convention).[117] Overall, Japanese officials are distrustful of China's intentions in space and believe greater transparency is needed to avoid problems. At the same time, despite Tokyo's participation in U.S. missile-defense developments and vague talk about possible future space defenses, there is a widespread belief that the country will not deploy space-based weapons or become involved in an arms race in space with its neighbors. Some officials believe such a course is impossible, given the country's peace constitution and Japan's long-established policies at the United Nations in support of the resolution on the Prevention of an Arms Race in Outer Space.[118] Others refer to Japan's alliance with the United States and argue that the benefit of these ties means that "Japan does not have to have an arms race" in space in order to maintain its security.[119]

Overall, Japan has used the Basic Space Law to move from a policy of exclusively "nonmilitary" uses of space to one of "nonaggressive" uses of space that can include a military component. However, there are differing opinions on whether Japan's Basic Space Law will lead to a more aggressive stance in space. Space alone will not determine the answer, however, which will be affected instead by Japan's overall security stance. In this regard, Kenneth Pyle argues that the emerging generation of Japanese politicians born and raised after World War II are "impatient with the slow pace of change in economic restructuring, in developing a more assertive foreign policy, and in rethinking the [peace] constitution."[120] Pyle argues that this postwar Heisei generation will be more self-assertive than its elders, who were constrained by the burdens of history. But the Japanese analyst Yoshinobu Yamamoto sees a more limited shift. He describes Japan's cautious remilitarization and increased participation in international peace-keeping and other operations as "activism lite."[121] In other words, he sees them falling short of any independent use of force. The Indian observer Bhubhindar Singh agrees, arguing that Japan has dropped its initial postwar "peace-state" identity merely in favor of a post–cold war "international-state" identity linked closely to UN-sanctioned actions in a multinational context. He believes that it will continue to reject the unilateral use of force or even collective self-defense with the United States.[122] But the issue of missile defense may begin to blur this line, as some Japanese experts and politicians around the time of the April 2009 failed North Korean missile/ space-launch test spoke openly of the need to work collaboratively with the United States against threats to either country from North Korea, because Pyongyang's ultimate intentions in a missile attack would be difficult to ascertain before a decision to intercept the missile would have to be made.

SPACE, REGIONAL PRESTIGE, AND COOPERATION

Japan is clearly pursuing a soft-power strategy in regard to space, hoping to use its accomplishments, unique technical capabilities, and outreach as a means of building respect and political influence within Asia and the world. This can be seen in its expanded efforts at space cooperation and the series of forward-leaning scientific missions it has on its agenda to accomplish space "firsts."

Japan's activities within the framework of the APRSAF organization provide solid evidence of the country's interest in serving as a leader in regional space activities. Japan has organized eighteen major conferences of this group, which aims to serve as a forum for space science research and applications for Asian countries. What is notable about the APRSAF's recent meetings is the gradual shift in venue away from Tokyo to other Asian locales where Japan can "show the flag" and promote new forms of space cooperation. Accordingly, recent meetings have been held in Jakarta (Indonesia), Hanoi (Vietnam), Bangalore (India), Bangkok (Thailand), and Melbourne (Australia). Its work covers a broad range of topics, including regular working groups on Earth observation, *ISS* utilization, communi- cations-satellite applications, and space education and awareness.[123] Since 2006, Japan has also used APRSAF to build a regional, space-assisted di- saster-management system, with separate meetings held throughout Asia. This so-called Sentinel Asia project has succeeded in enlisting participat- ing organizations on its Joint Project Team ranging from the Australian Bureau of Meteorology to the Bangladesh Space Research and Remote Sensing Organization to the National Disaster Reduction Center of China (People's Republic of China) to the National Remote Sensing Center of

Mongolia to the (South) Korea Aerospace Research Institute, although most countries (including China) are data recipients only, not providers.[124] Given the large number of recent natural disasters in Asia and the vulnerability of numerous countries to rising ocean levels and tsunamis, Japan's contribution of satellites, data-processing technology, and organizational support for Sentinel Asia marks some progress toward building regional trust.

Japan has historically sought to extend its regional influence through Official Development Assistance (ODA). A number of officials have identified space as an important new area for Japanese ODA.[125] Diet Member Kawai, for example, views space diplomacy as "the most important aspect" of Japan's space policy and argues that JAXA should provide launch access, satellites, and training to other countries in Asia via ODA.[126] To date, such efforts have had mixed success. In regard to Vietnam, for example, the Japanese government provided ODA for the purchase of a satellite. But, after considering various options, the Vietnamese eventually selected a French satellite. As one Japanese government official said philosophically, "In order to get one bird, we have to throw several stones."[127] Despite this failure, there seems to be a strong commitment to following this route until it eventually yields success, as Japan has done previously in other areas. JAXA's recent initiation of a joint Earth-observation and small multipurpose microsatellite design project with India, South Korea, Malaysia, Thailand, and Vietnam in the so-called Satellite Technology for the Asia-Pacific Region (STAR) project is one clear indicator.[128]

In the field of high-prestige space science missions, JAXA has charted a dynamic and ambitious course for its next decade and a half, showing its commitment to space but also its sense that it perhaps has something to prove. In the near term, Japan plans to return to the Moon in its *Kaguya-2* mission, with a lander for on-site experimentation.[129] It expects to take its next steps with humanlike robots and finally astronauts sometime after 2020. Japan sees itself as a key player in lunar exploration, although its means to access the Moon remain unclear. If the United States does not mount a major push, questions will arise as to whether Japan will join other players, such as China, Russia, or the European Space Agency. JAXA's "Roadmap to the Solar System" includes a string of planned missions aimed at orbiting Venus, a second asteroid mission, and probes to Mercury (with the European Space Agency) and Jupiter.[130] Japan is already a leader in space science research among its Asian neighbors and clearly has no plans of yielding the field to its rising competitors.

One Japanese official refers to China's success in using its spacewalk for soft-power purposes, making an interesting linkage to the role of space in helping to promote the country's international image as a technological leader and, down the line, China's potential ability to sell refrigerators, cars, and other products, especially in the Third World.[131] Japan, he argues, benefited similarly from its *Kaguya-1* lunar mission and needs to continue to conduct similar missions in the future to continue to promote its technological reputation.[132] JAXA's *Kibo* module has succeeded thus far in gaining a dozen government-to-government contracts with foreign space organizations to carry out experiments on the module, thus further building Japan's reputation as a regional and international space leader.[133] Such arguments are consistent with Pyle's observation that "The younger politicians are intent on diversifying the tools of Japanese foreign policy. They favor a higher profile for Japan."[134]

CONCLUSION

Japan's struggles in the late 1990s and early part of the twenty-first century in space have served as a "test" of its commitment. It seems clear that Japanese officials have decided that space is a signal activity for the country and requires additional political attention and budgetary support, not only for the country's continued development but for its military security and its overall "place" within the hierarchy of Asian nations. Japan is not simply standing aside and allowing China to pass it. Instead, it is continuing to capitalize on its lead in high technology to stake out a position in innovative space science, higher-end elements of the commercial market, and support capabilities in the military field (particularly communications and reconnaissance). Japan's leadership role in the APRSAF and its significant cooperation not only with the United States but also increasingly with India and South Korea indicate a soft-power strategy as well, perhaps as a counterbalance to China. Japan has extensive experience in such strategies at the commercial level and with industrial ODA. Whether such policies will be as effective regarding space remains to be seen. In any case, the continuation of its close relationship with the United States seems to signal

Japan's commitment to the next stage of lunar and possible Mars exploration, which could begin the steps to a permanent presence on those celestial bodies. Japan recognizes that it cannot do this alone. But doubts about the future U.S. astronaut-transportation system and a desire to increase its independence may cause Japan to move forward with the work (and risks) of human-rating its *HTV* module. This would put Japan in a better position to compete directly with China's space program, if it chooses to do so. On the other hand, Japan's style in space has not been confrontational, and it has sought to minimize losses rather than attempt overly risky gains. But Saadia Pekkanen's argument about Japan's past resiliency in the face of adversity is worth recalling: "The point is that they remain in the game for the widespread benefits, perceived and tangible, that are conferred upon the whole Japanese technological base and that affect Japan's national security in the long run."[135] Overall, Japan has the know-how, resources, and commitment to remain a formidable competitor in Asia's continuing space race.

CHAPTER THREE

THE CHINESE SPACE PROGRAM

From Turbulent Past to Promising Future

It is no exaggeration to say that China's space program, with in its rapid entry into human spaceflight and its equally demonstrative forays into military space projects, has captured world attention. Yet the origins and evolution of the Chinese space program—with its major discontinuities and changes in direction—are different from the fairly linear trajectories of most major space programs. In the United States, France, Japan, and the Soviet Union (at least after the 1930s), the pace of advancement tended to hinge mostly on questions of technology and funding. In China, by contrast, politics played a much more central role, often *constraining* the space program's development or causing radical shifts in its direction and purposes. As the top Communist Party leadership lurched from backyard industrialization in the 1950s to the Cultural Revolution of the 1960s, China's early efforts in space stagnated. Thus, China did not become a factor in the cold war's space race and ended up depriving itself of some of the benefits that the earlier acquisition of space technology would have provided in the fields of communications, agriculture, weather forecasting, and military reconnaissance. Yet Chinese work on missiles to deliver nuclear weapons continued unabated and facilitated China's first satellite launch in 1970. But only years later did a full-fledged space program emerge.

Conditions for space technology development changed dramatically after the fall of the radical Gang of Four in the mid-1970s and, particularly, after the rise of proreform policies under Deng Xiaoping in the late 1970s. Within thirty years—thanks to hard work, reliable state support, and the advantages provided by available foreign technology and know-how— China's space program has moved from a backwater to a leadership position within Asia, particularly in military capabilities and in independent human spaceflight.

Today, China has the second largest space budget in Asia (surpassed only by Japan), and that budget continues to grow, even as other programs in the region have slowed due to ongoing effects of the global economic downturn of 2008–2009. It has invested heavily in space infrastructure and in the training of personnel as well as in foreign outreach for both technology acquisition and for technology sharing with less-developed space actors. But questions abound regarding its ultimate motivations and direction. Some analysts see the program as driven primarily by military motivations.[1] Larry Wortzel argues, "it is clear that the PLA [People's Liberation Army] is serious about space warfare."[2] Other experts point instead to China's major focus on human spaceflight, the United States' vastly larger expenditures on military space, and China's extensive cooperative programs with other countries. For example, Joan Johnson-Freese argues that China is largely following a soft-power approach to space, relying on prestige, international cooperation, and commerce, plus modest military hedging.[3] Still others see China's space efforts as prestige oriented and best understood as part of the country's long, historical struggle to return to great-power status. Fiona Cunningham argues that China is plagued by a "sense of entitlement and victimization" dating back to the Opium Wars of the mid-1800s and that "Status is the most important motivation for a manned space program in the eyes of elite political leaders."[4]

Given these different interpretations, the politics of China's space program are worth examining in detail. Observers from all perspectives criticize Beijing for the lack of transparency in its space decision making. These problems seemed to plague even China's own Foreign Ministry in the immediate aftermath of the January 2007 antisatellite test: it first denied the action and then struggled to explain this sharp turn from China's erstwhile policy of opposing the "weaponization" of space. The China National Space Administration has also come into apparent conflict at times with the People's Liberation Army over policy priorities. In the Western analytical literature on Chinese politics, this phenomenon has been referred to as "fragmented authoritarianism," describing conditions in which rival groups of leaders—often institutionally based and organized—have overlapping power over a particular policy realm.[5] For these reasons, China's less-than-open space decision-making processes raise questions both about near-term priorities for the space program and about its longer-term direction and control. The result has been a widespread tendency in the U.S., Indian, and Japanese defense communities to react to this uncertainty by assuming the worst-case scenario.

Another reason that China confounds analysts is the presence of seemingly contradictory tendencies within the space program itself. As noted above, despite its military programs, China has long been engaged in international space cooperation and continues to expand these activities. It worked previously with NASA from the 1980s to the late 1990s and has cooperated with the European Space Agency (ESA) continually since the late 1980s. Similarly, following the normalization of Sino-Soviet ties in 1989 after decades of bilateral hostility, China purchased significant amounts of technology from the Soviet and later Russian space program and continues to work with Russia's national space agency on a variety of scientific projects. Finally, China is emerging as a key provider of technology and training to countries in the developing world, both through bilateral contacts and the Chinese-led Asia-Pacific Space Cooperation Organization, which China has visions of turning into an ESA-like organization. For these reasons, it is a mistake to view China as a closed monolith in regard to space, given its extensive international contacts and cooperative projects. Indeed, like the U.S. space program, the breadth of the Chinese space program and the presence of multiple factors behind its development and current activities make simple explanations of its status, current dynamics, and future direction difficult. It is also likely that China's path in space will be affected by the behavior of other countries.

With these complexities in mind, this chapter will describe, explain, and sort out the sometimes troubled route China's space program has taken from its infancy to its recent emergence as a major space power. It begins with the roots of the program in the mid-1950s, its travails in the 1960s to the late 1970s, and then its rapid rise since the 1980s. While China touts its independence in space, this situation emerged only after long years of assistance and technology purchases from other countries.

To some degree, China continues to seek outside help as it attempts to develop and deploy a broader range of space technologies. In this sense, it shares a variety of characteristics with most other late-developing space programs. Where China's space program differs is in the rate of its growth and the sheer scale of the resources available to it, which have outstripped all of its Asian rivals except Japan, moving it closer to the full range of space activities conducted by such leading powers as Russia, ESA, and the United States.

ORIGINS OF THE CHINESE SPACE PROGRAM

The roots of China's space program are humble. Unlike Germany and the United States and, somewhat later, Russia and Japan, China failed to industrialize and remained an agricultural country well into the twentieth century. It also lacked the advanced educational foundation needed for rocket and eventual space research. Chinese cultural factors periodically caused government bureaucrats to distrust foreign science, and prevailing conditions of foreign intervention, economic distress, and political turmoil hampered indigenous scientific development from the beginning of the twentieth century through the early post–World War II period.

Modern Chinese scientific organizations began to be constituted only in the 1920s in such fields as astronomy, physics, and chemistry. In 1928, the Nationalist government in Nanjing founded the Academia Sinica and began to establish a network of ten research institutes to promote Chinese science.[6] But the Japanese invasion of Manchuria in 1931 once again disrupted China's social and political order, although some scientists rallied to the side of the Nationalist government in the face of the Japanese threat.[7] The problem of scarce resources and increasingly unstable political, economic, and security conditions caused these efforts to fail. Many of the top Chinese scientists moved abroad during this period in search of more beneficial conditions for both higher education and research.

Unlike in Korea, where the Japanese occupation incorporated the population (often by force) into the empire's relatively advanced educational system, imperial Japan treated the Chinese as a foreign, subjugated people. Thus, poor conditions for scientific education prevailed throughout the

1930s until 1945, only to be replaced by civil war between the nationalists and communists. Finally, with the Chinese Communist Party's victory in 1949, China's science community began to have a chance to regroup, although under radically different circumstances than had prevailed in the 1920s.

The Communist Revolution and Scientific Reorganization

Chinese scientists now had to rebuild their research and educational institutions largely from scratch, facing constraints imposed by domestic communist ideology and under pressure to adopt the scientific structures established by China's much more advanced communist ally, the Soviet Union. In 1949, the government established the Chinese Academy of Sciences to serve as a central hub for technically oriented manpower and research, but its resources remained minimal. As Roger Handberg and Zhen Li estimate, "By 1949, China nominally had about 50,000 trained technical personnel, but only 500 of them were real researchers. Modern science and technology only existed in limited areas such as geology, biology, meteorology and fields not requiring experiments."[8]

The Beijing government, however, had one potential advantage in the large Chinese diaspora abroad, which included highly accomplished scientists, if only they could be attracted back to their homeland. In the space field, a steady trickle of experts began to filter back to the mainland after 1945 from the United States and the United Kingdom.[9] These individuals would eventually prove critical to the space program.

China's new alliance with the Soviet Union meant that it could both apply to Moscow for technical aid for the construction of research facilities and send its students to the Soviet Union for training, thus seemingly assuring that at least the next generation of Chinese would be technically skilled. In major industrial fields, China also benefited from large-scale Soviet investment and technology transfer, speeding development but also creating a dependency relationship. Yeu-Farn Wang recounts that during this period, "the Soviets delivered $3 billion worth of industrial equipment and machinery and helped China complete 130 projects, including factories for tractors, trucks, machine tools, and general equipment."[10] The Soviets eventually sent some 24,000 blueprints and other technical documents and 11,000 advisors to China.[11]

By the mid-1950s, some of the key building blocks had begun to fall into place for a rocket program. Cal Tech–trained Dr. Qian Xuesen, who had served in the U.S. military during World War II and participated in the removal of V-2 rockets and personnel from Nazi Germany, returned to China after being in and out of U.S. prisons and under surveillance since 1951 due to McCarthy-era charges of suspected communist sympathies. The United States deported Qian to China in 1955 with a group of other Chinese scientists as part of an exchange for American prisoners from the Korean War.[12] In the words of former U.S. Secretary of the Navy (1951–1953) Dan Kimball, "It was the stupidest thing this country ever did."[13] A few scientists would eventually join Qian in missile work, while others would end up in the nuclear program or in other areas where their advanced technical skills could be put to use. Organizationally, China's Central Military Commission formed the Fifth Academy under the Defense Ministry to lead military research and development on key strategic tasks facing the new nation, such as developing nuclear weapons against the perceived threat from the United States.[14] In 1956, senior officials tasked the Fifth Academy with development of missile-delivery systems as well, and the new body eventually moved under the new National Defense Science, Technology, and Industry Commission headed by Marshall Nie Rongzhen.[15]

Although political relations with the Soviet Union began to strain after Nikita Khrushchev's anti-Stalinist secret speech at the Twentieth Communist Party Congress in 1956, Moscow went forward with the delivery to China of two Soviet V-2-type R-1 rockets. This technology was already dated, however, and lagged considerably behind contemporary Soviet rocket technology, which would soon yield the world's first intercontinental ballistic missile (the R-7) in August 1957. The Chinese requested more advanced designs. After a planning process involving steady consultations with Soviet advisors, the Chinese government released a "Twelve-Year Plan for the Development of Science and Technology from 1956 to 1967," which included missile research as a priority area.[16] Moscow sent two Soviet R-2 rockets and a full set of design schemes for production and testing.[17] With these materials, the Fifth Academy initiated Project 1059, with the aim of copying the Soviet R-2 rocket and building it in China.[18]

A GREAT LEAP FORWARD INTO SPACE?

After the Soviet Union's successful launch of *Sputnik I* in October 1957, Mao Zedong also began to aim at launching a satellite. Showing typical revolutionary zeal but little practical knowledge, Mao exclaimed to Chinese scientists reporting on satellite technology before the Chinese Communist Party (CCP) plenary in May 1958: "If we're going to throw one up there[,] then throw a big one, one that weighs two tons."[19] Such a satellite would dwarf both *Sputnik* and the U.S. *Explorer I* satellite launched in January 1958 and require a booster larger than even the Russian R-7. China could not hope to achieve such a technological leap. Meanwhile, the Soviet Union had begun to hedge on its relationship with Beijing. Wang notes that as a result of the CCP's increasingly radical political direction and Moscow's new financial troubles caused by the increased demands on its finances in Eastern Europe after the uprisings in Poland and Hungary in 1956, "After 1957, Soviet exports to China were on a cash-and-carry basis, a step which placed the Chinese in an increasingly awkward economic position."[20] China's Great Leap Forward caused additional hardships, as food became scarce and critical production-related resources began to dry up too.

Nevertheless, the country surged to meet the goals of Mao's ambitious effort in the Great Leap Forward to industrialize by sheer willpower and the wide-scale application of utopian technologies (such as the ill-fated "backyard furnaces" campaign). Chinese scientists outdid one another in promising dramatic feats to lead these efforts. In this spirit, the leaders of the Chinese Academy of Sciences pledged to launch a satellite by October 1959.[21] China's nuclear and missile tsar, Marshall Nie—now head of both the Science and Technology Commission for National Defense and the State Science and Technology Commission[22]—added satellite development into the research program for the military's Fifth Academy (with Qian as deputy director) and the Chinese Academy of Sciences, under the new Project 581.[23] But none of this was to be, as China still lacked a launch vehicle and could not have produced one in time, even under favorable conditions. The Great Leap Forward soon led to massive starvation and major economic reversals rather than the progress its advocates had promised.

Still, rocket work continued, somewhat isolated from the tumultuous events taking place in the rest of Chinese society. Beyond the Project 1059 team working on the Soviet R-2 copy, scientists at the Shanghai Institute of

Machine and Electricity Design had started an indigenous effort in 1959 to develop a sounding rocket for use in meteorology and other scientific research.[24] Their initial design proved unsuccessful, however, in part from difficulties in obtaining critical materials needed for the use of planned cryogenic (liquid oxygen) fuel. Eventually, the so-called T-7M rocket—a two-stage liquid- and solid-fuel design—succeeded in reaching an altitude of eight kilometers from a launch site on the East China Sea at Laogang near Shanghai,[25] but this was far short of the over one hundred miles of altitude and much higher velocity needed to achieve orbit. Plans for a satellite quietly moved from a government priority to a backburner program in the Chinese Academy of Sciences.

By 1960, the critical technological relationship with Moscow had begun to fall apart over serious differences in political direction. China's radicalism and willingness to foment world revolution in the nuclear age had crossed swords with Nikita Khrushchev's vision of "peaceful coexistence" with the West and his desire for de-Stalinization and stability for the purposes of domestic economic development. Soviet advisors were pulled from China and technical assistance ground to a halt. As Wang summarizes: "Left with only prototypes, the Chinese had no choice but to reverse-engineer them, a process that took years."[26] This final break with the Soviet Union marked the beginning of a period "in the wilderness" for China's space scientists. Top-level support for space research soon faded, as in late 1962 Chairman Mao—in contrast to space backers President of the National People's Congress Liu Shaoqi and State Council Premier (Prime Minister) Zhou Enlai[27]—began to reemphasize the importance of "class struggle" in maintaining the momentum of the revolution. By 1964, Mao and his radical supporters began to characterize scientific institutes and universities as havens of bourgeois ideology.[28]

The one exception remained missile development. The emerging dual threats of Moscow and Washington as adversaries heightened China's need to construct a missile capable of delivering China's future nuclear deterrent. But Chinese missile engineers had only racked up a record of failure in their efforts to build and launch a copy of the Soviet R-2. In theory, the missile had a range of about six hundred kilometers, although without the payload capacity to carry a first-generation nuclear weapon,[29] which China had not yet built or tested. The Fifth Academy finally succeeded in launching the first such copy, renamed the Dong Feng (East

Wind, or DF)-1 in November 1960, from the military's new Jiuquan launch site in the Gobi Desert.[30] By 1961, some 15,000 Chinese personnel would be working on the missile effort.[31]

But the Fifth Academy's hopes of moving quickly to an intercontinental ballistic missile (ICBM) had to be curtailed until shorter-range systems could be developed, tested, and reliably built in a step-by-step fashion. Engineers succeeded in June 1964 in launching a 1,500-kilometer-range missile, the DF-2, capable of hitting Japan and U.S. forces stationed there.[32] By October 1964, China had tested its first nuclear weapon at Lop Nor. Space now moved under the so-called Seventh Ministry as part of a general reorganization of national security functions.[33] While China pledged never to be the first to use a nuclear weapon, it continued to work toward a missile capability to enable it to hit both the Soviet Union and the United States. Chinese Academy of Sciences personnel and Qian in the Fifth Academy now used the DF-2's success to attempt a revival of the national satellite program, with proposals brought before the CCP's Central Committee and the state Committee on Science and Technology for National Defense.[34] The proposals succeeded in gaining approval in 1965 for the launching of a hundred-kilogram satellite within six years.

In the meantime, the Chinese Academy of Sciences had begun to extend its sounding rocket program into the upper reaches of the atmosphere and to experiment with the launch and return of animal payloads such as mice, rats, and dogs. Using a modified T-7A booster, the Shanghai Machinery and Electrical Equipment Design Academy and the Academy of Sciences' Biophysics Research Institute cooperated to carry out a series of suborbital flights from 1964 to 1966.[35] This research paralleled earlier studies on animals conducted in the Soviet and U.S. space programs, with the ultimate aim of launching humans into space. But the start of Mao's Cultural Revolution in 1966 intervened, halting this line of research and forcing many of the top scientists out of the program.

SPACE UNDER THE CULTURAL REVOLUTION

Mao's Cultural Revolution sought to disrupt the bureaucratization of Chinese politics, mobilize China's youth to participate in revolutionary activities, and root out the appearance of a new, technically focused bourgeoisie within Chinese society.[36] The move also sought to dislodge factions surrounding President Liu Shaoqi and General Secretary Deng Xiaoping, who threatened to eclipse Mao at the center of the CCP hierarchy, by fomenting a revolution from below. But Mao found his own role somewhat weakened as the new movement picked up steam. The evolving leadership under the later-named Gang of Four struck hard at the traditional power bases of its opponents: anyone with foreign (even communist) contacts, the industrial elite, and the educated classes. A Beijing Red Guard statement issued in 1967 outlined the range of enemies of the revolution: "First, we want to lodge the strongest protests against you . . . that serve the bourgeoisie . . . and the so-called technical authorities who want to take the capitalist road!"[37] Training of young specialists ground to a halt as the Gang of Four disbanded most universities and institutes in favor of practical agricultural experience, self-criticism, and a mass movement of "going to the people" to purify the revolution. The government halted publication of almost all technical journals by the end of 1966.[38] By the following year, an all-out attack on scientific research institutes was underway, with the new mantra of "self-reliance" being proselytized to strengthen the bonds between scientists and the masses. For those within the Academy of Sciences, the results of these attacks came swiftly and with devastating effect on their research programs. During the next decade, the academy suffered a loss of five thousand personnel and the slashing of its affiliated institutes from 106 to a mere forty-one.[39] Qian found himself stripped of his leadership role in early 1967 and forced to sign a confession; other senior scientists ended up in the countryside, "locked up in cattle sheds,"[40] or dead.

Premier Zhou Enlai sought to protect a small number of top rocket scientists, who continued to work on high-priority projects,[41] such as military missiles and China's first satellite launch, out of a need for military reconnaissance. In 1968, launch-vehicle research moved into the Chinese Academy of Launcher Technology (CALT), and a number of satellite-related institutes were consolidated into the Chinese Academy of Space Technology (CAST).[42] General Lin Biao—an increasingly important political player by the late 1960s—emerged as a key force in maintaining the space program, whose technology he recognized would be necessary for strengthening and modernizing the PLA. In the midst of this turmoil, remarkably, the National Defense Science Committee set up a special institute in 1968 to begin a process of training and selecting astronauts from the Chinese military, with a planned launch date of 1973.[43]

The military's drive to develop a nuclear weapons delivery system led to the deployment of the liquid-fuel DF-3 and then longer-range DF-4 missiles. Soon, CALT was assigned the task of adapting the DF-4 for a satellite launch, by adding a solid-fuel third stage.[44] The work proved difficult, however, given China's lack of expertise in solid-fuel technology and the absence of foreign assistance. Using the T-7 sounding rocket as a test vehicle, CALT engineers eventually succeeded in conducting a test of the third-stage engine in August 1968, following with a static test of the much larger DF-4-derived rocket, now called the Long March 1 (or Chang Zheng), the following summer.[45]

Chinese technicians gathered the various stages at the Jinquan launch site in early 1970 along with a simple *Dong Fang Hong-1* (*DFH-1*) satellite, weighing in at 173 kilograms (much larger than the eighty-four-kilogram Soviet *Sputnik-1* in 1957 or the fourteen-kilogram *U.S. Explorer-1* in 1958 but far short of Mao's desired two tons).[46] But would the system work? With Premier Zhou Enlai on site, the three-stage rocket blasted off on April 1, 1970, and succeeded in placing the *DFH-1* into orbit, making China the fifth country to accomplish such a feat, just a few months after Japan. The satellite underlined China's accomplishment with a signal that beeped the melody of "The East Is Red" for twenty-eight days, until its batteries gave out.[47] The success of the launch led to plans for additional satellites. China also moved forward with the selection process for its first astronauts.

In October 1970, the recently formed National Space Medicine Institute in Beijing collected an initial group of 1,840 pilots from the air force to begin testing.[48] After surviving a rigorous, ten-step winnowing process, the selection board nominated nineteen astronaut candidates in March 1971.[49] That same month, the Chinese military succeeded in launching China's second satellite (*Shi-Jian-1*, or *SJ-1*), which was equipped with solar cells capable of powering a multiple-year scientific mission. It carried a magnetometer, two detectors for x-rays and cosmic rays, a thermal regulating system, and a radio transmitter.[50] The *SJ-1* would function for eight years. China's space scientists seemed to be emerging from the dark years of the Cultural Revolution. As Thomas Fingar describes: "Beginning in 1971 there was a pronounced move to rehabilitate scientists, engineers, and other technical personnel discredited during the Cultural Revolution."[51]

Despite these triumphs and a new degree of acceptance, Chinese space scientists saw their lofty plans put at risk in September with the sudden death of their patron, Lin Biao, in an airplane crash over Mongolia, purportedly en route to the Soviet Union after a failed coup attempt against Chairman Mao.[52] This event would mark the cresting of the military's role in politics and caused a purge of his closest associates in the space program.[53] As Handberg and Li write, "The ambitious plans made in August 1970 to develop eight new launch vehicles and fourteen new satellites in five years disappeared."[54] The astronauts, similarly, soon found themselves without a program, as human spaceflight fell victim to a renewed political struggle.[55] The next satellite launch would not occur until 1975.

Exposure to Western technology and scientific accomplishments after President Richard Nixon's trip to China in 1972, however, highlighted to the CCP's leadership just how far the country was falling behind the rest of the world as it sought to purify its revolution. Chinese aerospace scientists were allowed to attend the Fourth International Aeronautics and Astronautics Exhibition in Japan in 1973 to begin studying the problem.[56] More importantly, China invited several groups of American scientists for exchanges in 1974, including visits by a number of senior physicists from major U.S. universities and governmental laboratories.[57] The information Chinese scientists and their leadership gained must have been humbling, exposing severe gaps in the now sharply curtailed Chinese system of higher scientific education and problems in the politicized framework for advanced research.

Some work continued on a more advanced launch vehicle for the space program called the Long March 2. Mao's elevation of Zhou Enlai's protégé Deng Xiaoping and his eventual decision to castigate the so-called Gang of Four in 1975 marked a decisive shift toward greater political pragmatism. That same year, China launched a large, two-ton recoverable satellite aboard the Long March 2 to help pave the way for military reconnaissance from space.[58] The top leadership also approved plans for development of a larger booster to facilitate the launch of satellites to geostationary orbit in order to develop satellite communications.

But, in politics, Deng's "rightist" (pro-Western) tendencies led to his temporary fall from favor and Mao's elevation of Hua Guofeng as his heir apparent.[59] Despite this attack on Deng, his senior ally Zhou Enlai outlined a new set of national priorities at the January 1975 National People's

Congress, calling China to rally behind the so-called four modernizations: the goal being to catch up with the West in agriculture, industry, defense, and science and technology by the year 2000.[60]

In the space field, although the Shanghai group had succeeded in launching several satellites in 1975 and 1976 and recovering their intact capsules, the technology for taking photographs and returning the images intact to Earth still had not yet been perfected. As Johnson-Freese characterizes the situation: "remote sensing was a new field of space applications . . . and with the educational gap left by the Cultural Revolution, there were few in China with any exposure let alone expertise in the field."[61] In addition, while China had contacted the International Telecommunications Union about its intention to place a satellite into geostationary orbit in 1975,[62] it still lacked both a capable booster and the necessary electronics to build and operate such a satellite.

THE RISE OF DENG AND REFORMS IN SCIENCE POLICY

Zhou Enlai's death in January 1976 and the momentous passing of Chairman Mao in September set up a final struggle between the two main political factions. Deng eventually succeeded in outflanking Mao's chosen successor, Chairman Hua, and assumed effective control of the reins of power by the end of 1978. In one of the key areas affecting science—education—Deng's initial policies marked significant changes from the Cultural Revolution's focus on political criteria, establishing instead new merit-based criteria in scientific education.[63] As one 1980 analysis of the problems in Chinese science summarized the leadership's challenge: "The manpower question is . . . perhaps the major constraint on scientific development."[64] But although the policies outlined by Deng began to put the space program on track, it would take time to make up for the lost years of the Cultural Revolution.

With China at risk of permanent underdevelopment in science and technology compared to its two major rivals (the United States and the Soviet Union) and regional rivals such as Japan, Deng argued successfully to the CCP that China had put its security at risk. This priority required the reestablishment of professionalization and a downplaying of politics. As Richard Suttmeier summarizes, "Under the new policies . . . professional norms and values were legitimated to an extent not seen since the mid-1950s."[65] However, as Wang explains, the CCP also "realized that the S&T [science and technology] sector . . . could not be left to its own devices."[66] With Deng moving to modernize China's economy and remove the excesses of the Cultural Revolution, space found itself recast as a leading edge of technological development that would help "pull" the country forward. The priority thus became the use of space for economic development through the application of communications and remote sensing rather than sophisticated military assets,[67] which were still out of its reach. Even remote sensing remained rudimentary. A Chinese Academy of Scientists representative admitted in 1982 at a Tokyo conference that the country's Earth-resources satellite was still stuck in "the initial designing phase."[68]

Emerging Ties with the United States

The Chinese government became more active in trying to jumpstart development by sending more scientists abroad to international meetings and allowing more visits by foreign scientists. Dr. Frank Press, who had traveled to China with other physicists in 1974, led a delegation of major U.S. federal scientific agencies to China in 1978 as President Jimmy Carter's science advisor.[69] These developments helped spur the integration of China's space scientists into the broader realm of international scientific exchange and set the groundwork for talks with NASA. In the short term, however, China needed critical technologies. As part of a general (but still temporary) reversal of 1960s-era policies of economic and industrial self-reliance, the government used Press's visit and negotiations on the normalization of official diplomatic relations to approach the United States about the purchase of an advanced communications satellite.[70] China also moved forward in negotiations with NASA on the possible U.S. hosting of Chinese scientific payloads on future missions on the planned space shuttle. Following a late 1978 visit to the United States by Ren Xinmin, the director of the Chinese Academy of Space Technology (who had received his Ph.D. from the University of Michigan and returned to China in 1949), the two sides reached an Understanding on Cooperation in Space Technology, which established a joint commission and several working groups.[71] The pact called for U.S. assistance in developing a civil communications and broadcast system for China through the purchase of a U.S. satellite, which would be launched by NASA but operated afterward by the Chinese.[72] (However, problems with financing on the Chinese side and the absence of a government-to-government agreement led to the breakdown of the proposed deal.) In addition, the agreement called for the Chinese purchase of a U.S. ground station for accessing Landsat remote-sensing data. The new relationship helped allow Vice Premier Deng Xiaoping to visit the Johnson Space Center in Houston in 1979, where he operated the controls on a simulator of the planned U.S. space shuttle.[73]

Given shared perceptions of the Soviet military threat, the two sides also began to cooperate more extensively after the victory of Ronald Reagan in the 1980 election, which created a top-level strategic rationale below which contacts in at least space science research could be justified and maintained. In 1984, as the Soviet Union began to offer guest cosmonaut flights for its allies to its *Salyut* and later *Mir* space stations, President Reagan offered a slot on a future U.S. shuttle flight to a Chinese astronaut.[74] While the manned mission never occurred (in part because of the *Challenger* disaster in 1986), cooperative scientific exchanges eventually came to fruition in January 1992, when two Chinese experiments flew aboard the U.S. space shuttle.

China's new push to develop space technology was rooted in a national plan for scientific and technological development for the 1978–1985 period.[75] The plan referred to comparative space developments in other countries and listed Chinese priorities for the coming period as: satellites for remote sensing, ground stations, space science research, "skylabs," and new launch vehicles.[76] The government also established a Ministry of the Space Industry in 1982 to coordinate these activities and to become the new public face of the space program.[77]

Given China's failure to acquire a U.S. communications satellite, however, the space program was forced to look elsewhere. In April 1978, China reached an agreement with France and West Germany to use a communications satellite already in orbit and later established an agreement with the West German Ministry of Research and Technology for cooperation in various fields, including space.[78] The Chinese also continued domestic development work. Led by Ren Xinmin,[79] the communications satellite program persevered through a series of difficult test problems.[80] In January 1984, from China's new, more southerly launch site at Xichang, the team managed to place a satellite into a low elliptical orbit with a partially failed Long March 3 booster.[81] But the satellite's problematic orbital location required more station keeping than could be accommodated with the on-board fuel supply. China finally achieved its first successful insertion into geostationary orbit that April, allowing nationwide television broadcasts for the first time on an experimental basis. After the completion of these trials, its first fully functional domestic communications satellite (the *DFH-2*) reached orbit and began operation in February 1986.[82]

Meanwhile, ties with the United States began to be forged in the area of commercial space, as China sought to enter the international launch-services market. Formed in 1980, the Great Wall Industry Corporation (GWIC) had originally been organized to try to import production technology from Japan and Hong Kong.[83] But in 1985, the government decided to task GWIC with marketing the Long March 3's launch services, based on its twelve successful flights. In 1988, President Reagan agreed to allow U.S.-made satellites to be launched on Chinese rockets. The deal required China to sign technological safeguard and liability agreements, establishing yearly government-to-government meetings to share views and discuss any problems. China had already joined the Outer Space Treaty in 1983, but its new activities now caused it to move forward in December 1988 with ratification of both the 1972 Convention on International Liability for Damage Caused by Space Objects and the 1975 UN Registration Convention. These steps marked a major shift in China's integration into the world space community and its growing acceptance of international norms.

Soon, GWIC had opened offices in Washington, in New York City's World Trade Center, and, in 1989, in Torrance, California, to be closer to U.S. satellite manufacturers who might purchase its services.[84] GWIC landed its first two contracts with the Hughes corporation, one for the launch of a satellite built for Austrian Telecom and another for *Asiasat-1*, which ended up being launched in April 1990 into geostationary orbit on a Long March 3 at a price of about $39 million.[85] Despite the political problems in U.S.-Chinese relations caused by the crackdown at Tiananmen Square in 1989, President George H. W. Bush waived economic sanctions to allow the launches to proceed. Other companies seeking to benefit from the Long March's low price and growing record for reliability followed, including Loral, Martin Marietta, Intelsat, and Echostar.

Meanwhile, China continued to develop its own satellite program—launching nine satellites from 1975 to 1987—and working on technologies associated with reentry and recovery of spacecraft.[86] These missions helped China gradually restore and expand its capabilities and put it on the verge of serious research into human spaceflight. At the same time, reforms in the science and technology sector had been incomplete. As one observer concluded at the time regarding the 1980s, despite the institutionalization of "modern science" as a national objective, "political and economic institutions capable of promoting technological advance remain backward and ineffective."[87] In space, this responsibility rested with the still military-led Ministry of the Space Industry.

As it sought to establish itself as a capable space power, China conducted six flights of its more advanced Fanhui Shi Weixing (FSW) series spacecraft from 1982 to 1987.[88] These flights carried microbiology payloads into space for testing applications of zero gravity for plant growth and crystal development. The *FSW-9* flight in 1987 carried two payloads for a French company, representing one of China's first commercial contracts. The FSW flights also experimented with photographic remote sensing, returning wide-angle images from film returned to Earth suitable for basic land use and navigational surveys but far from military standards.

Military Concerns

In the defense sector, China reacted with unease to the Reagan administration's plan in 1983 to develop a space-based Strategic Defense Initiative (SDI) for defending itself against Soviet missiles. While not aimed at China, the leadership decided the issue needed to be investigated as a possible threat in terms of technology.[89] China also faced a broader question as to whether its current strategy of agricultural reform and light-industry development would be adequate to meet the challenges of the emerging international environment. In April 1986, at perhaps the height of the power of the traditional military-industrial complex, a group of two hundred military and civilian experts convened by the Commission on Science, Technology, and Industry for National Defense (COSTIND), the Science and Technology Working Group of the State Council, and the State Science Commission issued a report to consider a possible high-technology strategy for the country.[90] In the face of a set of divided recommendations as to whether to pursue a military or civilian direction, Deng Xiaoping weighed in on the side of a plan for developing dual-use technologies, listing seven priority areas, including aerospace.[91] Space was soon subsumed under COSTIND.[92] According to Chinese scientists, concerns about SDI and the 1985 U.S. antisatellite (ASAT) test in space prompted the creation of a research program to examine similar technologies as a potential "gap" area for China.[93] In regard to civil space, a space committee meeting in 1987 sketched out specific plans for high-prestige human spaceflight operations in low-Earth orbit as one desired national objective for the early twenty-first century.

One of the key goals of earlier research into recoverable capsules had been the development of military reconnaissance capability, a critical function mastered first by the United States in 1960 and by the Soviet Union in 1962. Since then, those countries had advanced well beyond photographic images and film-return capsules to radio waves and then digital signals to provide real-time imaging, all-weather radar, and infrared capabilities. By contrast, China lagged considerably, thus greatly limiting its ability to conduct reconnaissance on other countries' military capabilities and carry out economically beneficial remote sensing of its own territory. Although China's FSW series and later *Feng Yun* (Wind and Cloud, or *FY*)-1 meteorological satellites represented steady improvements, as the military expert James Lewis argues, they were "still closer in capability to Landsat than to an intelligence satellite" in terms of their resolution.[94] In the late 1990s, China eventually worked with Brazil to develop a higher-resolution Earth-resources satellite (*Ziyuan-1*) capable of digital transmissions, but China still faced considerable limitations, forcing it to rely on purchases of foreign satellite imagery.[95]

Thanks to major changes in Soviet foreign and military policies under General Secretary Mikhail Gorbachev, Moscow and Beijing reestablished diplomatic relations in 1989. After the Soviet Union's break-up in December 1991, President Boris Yeltsin continued to court China as a market for Russian products and as a source for needed foodstuffs, textiles, and electronics. China, for its part, sought to renew trade and technology relationships in a number of fields that had been severed in the early 1960s. Space emerged as one of these areas. With China planning for eventual human spaceflight, Minister of National Defense Chi Haotian traveled to Russia's

Star City space-training facility in 1993 to investigate prospects of acquiring know-how and technology.[96]

The second-generation Chinese FSW spacecraft had a more advanced life-support system, which allowed China to launch and recover not just small plant and crystal samples but also conduct orbital experiments involving animals. To gain experience and hardware designs, China turned again to the Russian Space Agency, whose wares could be purchased cheaply. Following General Chi's trip, the two space agencies signed a formal space cooperation agreement in 1994. This led to a visit to Russia by a large Chinese space delegation the following year. The Chinese side purchased a spacesuit, a complete Soyuz capsule, docking equipment, a life-support system, and a variety of other hardware and design information to guide their planned human spaceflight program.[97] Chinese delegations returned in 1996 and 1997, gathering more information on cosmonaut training techniques and space medicine.[98] This Russian equipment and know-how, despite later Chinese comments about self-sufficiency in space, proved critical to the eventual success of the Shenzhou program. As described by Tai Ming Cheung, the Russian Soyuz proved particularly beneficial in "shortening the development cycle of the program and allowing the Chinese to make a generational skip."[99]

In order to facilitate its increasing cooperation with other countries in space and to put an ostensibly "civilian" face on an otherwise hidden, military-run space program, China created the China National Space Administration (CNSA) in 1993. However, while China portrayed the CNSA as its NASA equivalent, the bulk of China's space research, production, and operational functions remained outside of the new organization and within the defense industry. The main state-run enterprise group within the space sector had evolved from a series of military-led bodies: the Fifth Academy, the Seventh Ministry, the Ministry of the Space Industry, the Ministry of the Aerospace Industry, and, by the late 1990s, eventually the China Aerospace Corporation.[100] Finally, in July 1999, the State Council created a consolidated Chinese Aerospace Science and Technology Corporation (CASC), which brought together various research and production complexes as well as such organizations as GWIC.[101]

By the mid-1990s, China had made significant progress in space applications. Beijing reported that communications satellites now reached 83 percent of China's population, offering broadcasts, telephone communications to remote areas, data transfers (including printing national newspapers remotely), and educational services.[102] Finally, the Chinese people began to see evidence that the space program was having an effect on their daily lives.

NEW COMMERCIAL DEVELOPMENTS, SPACE OUTREACH, AND SETBACKS

China focused considerable attention in the 1990s on the development of a reliable booster for the international market, as well as its domestic space needs and expanding its space infrastructure. China's GWIC had succeeded in winning U.S. approval for launches in part because it had agreed to a quota system limiting the number of flights open to bidding on Western satellite launches and keeping the price within a certain percentage of average world launch costs, so as not to threaten American launchers unduly. Washington could set these terms because the vast majority of satellites in the world at that time had U.S. components, which were governed by U.S. export control regulations. The first quota, from 1988 to 1994, allowed China to contract for nine international launches.[103] A new agreement provided for an additional eleven launches from 1995 to 2001.[104] Through this quota system, China and a number of U.S. satellite companies developed regular business relations. Chinese and American officials also held regular meetings. By 1999, twenty-six U.S. satellites had been launched aboard Long March boosters (some in multiple-satellite launches).

But the United States was not China's only client. Pakistan contracted with GWIC to launch its first small satellite, *Badr*, into low-Earth orbit aboard a Long March 2E booster in July 1990, paying China about $395,000.[105] It also launched satellites for Sweden, Australia, and a variety of other Western customers. But China eventually experienced growing pains as it sought to establish itself as not only a cheap but also *reliable* provider of space launches. In 1992, a controversial incident led to the loss of a satellite after an explosion of the rocket's third stage, although China and a U.S. satellite provider disagreed about who was to blame for the failure. In 1995, a much more serious explosion of a Long March 2E booster and its payload occurred shortly into a launch involving a Hong Kong satellite, killing six people on the ground and injuring several dozen others.[106] China suffered a third launch failure soon after, when its new Long March

3B rocket in February 1996, carrying a payload for Intelsat, exploded even earlier in its flight; Chinese television interrupted its coverage rather than broadcast the disaster to its domestic audience. One foreign observer who visited the crash site estimated that some one hundred people died in a local village, although Chinese authorities later reported only a handful of deaths and some additional casualties.[107] Finally, a fourth launch problem occurred in August 1996 when a failed upper stage of a Long March 3 left a planned geostationary communications satellite built by the U.S. Hughes corporation for China in a useless low-Earth orbit.[108] The investigations of the Intelsat flight's explosion by the U.S. satellite firm Loral and of the August 1996 flight by Hughes experts and the sharing of their findings with Chinese space officials, would soon become a major bilateral incident amid U.S. government charges of improper export control compliance by American company representatives. China's less-than-open behavior regarding the investigations also did not help matters.

The Break with the United States and China's New Outreach Efforts

Spurred initially by fears of nuclear espionage, the congressionally mandated Cox Committee made a series of highly provocative charges in the missile and space field in the spring of 1999. Its report argued that the investigations conducted by Loral and Hughes had provided know-how of direct benefit to China's nuclear-delivery systems.[109] The specific issues involved satellite fairings that could be applied to warhead design and test diagnostics, although neither company was found guilty of transferring classified information. Both companies ended up paying fines for lesser charges having to do with unsupervised contact with Chinese space personnel during the postaccident investigation process. Critics of the Cox Committee report case raised doubts about its broad charges, arguing that the information was not sensitive and not likely to be particularly relevant to the military's missile programs.[110]

But the committee's findings succeeded in convincing other members of Congress to push legislation through both houses in 1999 requiring the Clinton administration to—among other restrictions—recategorize all space technology as munitions items under the U.S. International Traffic in Arms Regulations (ITAR). U.S. space cooperation with China ground to a halt. Through the rest of the Clinton administration and both terms of the George W. Bush administration, no U.S. satellite would be launched on a Chinese booster, and no meaningful cooperation in civil space with China would take place. The path of U.S.-Chinese space cooperation established under the Reagan administration had ended, and a period of enmity in regard to space ensued, supported initially by strong military and conservative political groupings on both sides.

Despite China's problems with the United States, it continued to work with other countries to acquire technology and know-how and to promote its space interests. In terms of technology development, China engaged in cooperation with Brazil, France, Russia, Ukraine, and the United Kingdom for collaborative missions or actual joint development of spacecraft.[111] For example, China sought experience in microsatellites through a partnership with Britain's Surrey Satellite Technology, Ltd., via a company affiliated with Tsinghua University. This cooperation yielded China's first microsatellite in June 2000, which entered orbit as a secondary payload aboard a Russian rocket. Tsinghua University began making images available publicly from its on-board camera system.[112] This company has since produced other satellites both with Surrey and other Chinese institutes.

In 1992, China began a series of space outreach meetings under the framework of the so-called Asia-Pacific Multilateral Cooperation in Space Technology and Applications (AP-MCSTA), with the aim of becoming a leader in regional space cooperation. In subsequent years, China continued to sponsor meetings of AP-MCSTA to exchange information and to begin planning for cooperative activities, holding conferences in Thailand, Pakistan, South Korea, Bahrain, Iran, and China.[113] AP-MCSTA set up a *Small Multi-Mission Satellite* (*SMMS*) project in 1998 as a joint hardware-development activity and proceeded with a series of training programs and cooperative exchanges.

Military Programs and Space Arms Control Initiatives

China had entered the 1990s with little discernable expertise in military space. During the next fifteen years, however, it began to build an infrastructure for military space operations. In the 1991 and 2003 Gulf wars, China witnessed the expert U.S. use of space and its critical role in large-scale, modern military operations. Given the military's refusal to cede space to the United States in any future conflict over Taiwan, Beijing moved to redress this imbalance. Building blocks in this process included more advanced reconnaissance and communications satellites, a sea- and dispersed ground-based space tracking network (which required deals with foreign countries to host Chinese ground stations), and continued work toward an ASAT weapon for use against foreign satellites in low-Earth orbit.

Meanwhile, China also began a policy of supporting new space arms control, perhaps as a hedge against the possible failure of its military systems or in an effort to halt a U.S. movement toward national missile defenses. Beijing opposed a U.S. withdrawal from the Anti-Ballistic Missile Treaty, which China believed could result in space-based interceptors that might pose a risk to China's limited nuclear forces.

In February 2000, China offered a working paper to the UN Conference on Disarmament (CD) in Geneva on suggestions for preventing an arms race in space. In the document, the Chinese Foreign Ministry offered a series of "tentative ideas," admitting the utility of military satellites ("their role should not be all together negated") but urging all countries "not to test, deploy or use any weapons, weapon systems or components in outer space."[114] The Chinese Foreign Ministry fleshed out these ideas in a 2001 working paper entitled "Possible Elements of the Future International Legal Instrument on the Prevention of the Weaponization of Outer Space."[115] The proposal outlined a broad agreement whose main points would obligate states not to test any weapons or components in space, not to test any such systems meant for space elsewhere, and not to launch objects into space "to directly participate in combatant activities."[116] In 2002, China joined with Russia to initiate a proposal at the CD on "Possible Elements for a Future International Legal Agreement on the Prevention of the Deployment of Weapons in Outer Space, the Threat or Use of Force Against Outer Space Objects." The document drew on an earlier Soviet treaty proposal from 1983 (seeking to block the U.S. Strategic Defense Initiative) with some elements of the Chinese working paper.[117] The draft treaty was the first significant new space arms control initiative in years and called simply for states "Not to place in orbit around the Earth any objects carrying any kinds of weapons" and "Not to resort to the threat or use of force against outer space objects." But the treaty offered no definition of "weapons" and failed to include the elements in China's 2001 draft that would have prohibited the testing of space-to-space, ground-to-space, or sea-to-space weapons. It also contained no specific verification measures. Because of the CD's failure to agree on an agenda, the proposal would languish for the next decade. Nevertheless, Chinese diplomats would continue to mention the proposal as an alternative to purported U.S. plans for "weaponizing" space.

BUILDING A MAJOR SPACE PROGRAM

The decade from 2000 to 2010 represented a "coming out" party for China's space program on many levels. Its human spaceflight program suddenly achieved an unexpected victory over Asia's more experienced space programs. Its military conducted a successful test of an ASAT weapon, shocking the world's most advanced space powers. And China's space industry registered dramatic improvements by offering more advanced satellites and launch vehicles and by reestablishing profitable contacts with foreign partners—and not only in the developing world. Advanced space countries (such as France and Britain) were now eager enough to work with China that they developed specific ITAR-free spacecraft to get around U.S. restrictions. Russia also continued to cooperate with China and provide technology and know-how. In many respects, Beijing had successfully outflanked the U.S. sanctions, although it also attracted unwanted attention to the continued, central role of the military in its space program.

Civil Space Activity

China's space science and exploration program had evoked little interest from the international community beyond space specialists until its October 2003 *Shenzhou V* launch. While experts had long expected the flight, after a series of unmanned missions with the Shenzhou capsule in previous years, the actual flight came as a jolt to the general world public and to political leaders. At the regional level, in particular, as Johnson-Freese recounts, "Asian governments and publics paid considerable attention."[118] China successfully followed its first one-man flight with a two-man flight in October 2005 on *Shenzhou VI* and then a three-man flight with a spacewalk on *Shenzhou VII* in October 2008. Chinese television proudly broadcast the 2008 mission and its spacewalk. But the flight caused its greatest stir abroad because of its release of a forty-kilogram picosat (*BanXing* [or *BX*]-*1*), which took pictures of the *Shenzhou VII* and ended up some twenty-five kilometers from the *International Space Station* (*ISS*).[119] While later reports downplayed the risk of a collision, foreign space analysts remained perplexed at China's apparent willingness to come that close to the *ISS*. Others argued that it represented a test of a military-purpose "killer satellite." However, its lack of any apparent explosive mechanism and the transparency of the mission—which was widely reported on in China—discounted this possibility, for at least the first *BX-1* flight.

In its 2006 White Paper on space, the Chinese government announced, "Having made a historic breakthrough in manned spaceflight, China has embarked on a comprehensive lunar exploration project."[120] This statement fueled speculation about China's intentions. China's first high-prestige space science mission came in the form of the *Chang'e 1* lunar probe, which orbited the Moon from 2008 to early 2009, mapping the lunar surface and analyzing the lunar environment.[121] The mission, while costing a reported $205 million, could not match the sophistication of Japan's *Kaguya* probe, but it helped establish China as a serious player in lunar science.

China continued to contract with Russian space enterprises for their expertise in instrumentation, equipment, and control systems for major space missions. In 2006, Russian Space Agency Deputy Director Yuri Nosenko reported that China signed contracts with Russian space enterprises worth tens of millions of dollars.[122] The two sides announced plans to cooperate on lunar- and Mars-related robotic exploration, including with automated rovers.

To cement its space leadership role within Asia, China led a subset of interested AP-MCSTA participants in creating a more formal, dues-paying membership body for Asian space activity in 2005: the Asia-Pacific Space Cooperation Organization (APSCO). Bangladesh, Indonesia, Iran, Mongolia, Pakistan, Peru, and Thailand joined China in the initial membership group of APSCO, which established its headquarters in Beijing. After the ratification of the protocols by the various national governments (minus Indonesia), APSCO was formally inaugurated in December 2008.[123] The fact that China was the only major Asian space power in APSCO (with even South Korea declining to join) put it in a clear leadership position, although the group's activities remained modest, focusing mostly on the training of foreign scientists at Chinese institutions (such as the Beijing University of Aeronautics and Astronautics) and the donation of ground stations to member countries to receive information from Chinese satellites. Despite initial plans for cooperative development of the *SMMS* satellite, Chinese institutes eventually built the spacecraft and donated it to APSCO to conduct disaster warning and environmental monitoring for participating countries.[124]

In terms of Sino-U.S. space relations, however, very little contact between the two space agencies took place. The U.S. Congress had weighed in after the Cox Committee report and made such contacts prohibitively complicated; it also specifically denied China access to the *ISS*. The first tentative step toward a rapprochement came in April 2006 when CNSA Vice Administrator Luo Ge visited the United States and invited a reciprocal visit by the NASA administrator. NASA's head, Michael Griffin, went to China in September 2006. However, he was denied access to the Chinese flight operations center and other requested locations. Part of the problem may have had to do with an internal Chinese dispute between the CNSA and the PLA over permission to visit these military-controlled facilities.[125] The visit failed to stimulate good feelings or strong ties, although two expert-level working groups on Earth and space sciences had some limited follow-up contacts for data exchanges.

During President Obama's summit meeting in Beijing in November 2009 with President Hu Jintao, the two leaders surprised many observers by specifically including a clause in their joint statement calling for "expanding discussions on space science cooperation and starting a dialogue on human space flight and space exploration."[126] It remained to be seen what substantive activities would come from this statement, given the continuing obstacles in the form of PLA and U.S. congressional opposition.

A surprise visit by NASA Administrator Charles Bolden to China in October 2010 with the announced plan of discussing possible cooperation in human spaceflight raised hackles in some quarters of Capitol Hill. Representative Frank Wolf, a Republican from Virginia, complained of Bolden's failure to coordinate with Congress on the trip and reminded the NASA administrator that "Several recent NASA authorization bills have explicitly sought to place restrictions on coordination with China."[127] Bolden reportedly received far better access to Chinese space facilities than was accorded to his predecessor. His host, however, was the Chinese military, although he noted that Beijing is "struggling right now with how they split up responsibility for [its space] programs."[128] No concrete results emerged, except plans for future meetings.

Military Space and Arms Control

Results from China's accelerated work on a number of military space programs since the early 1990s could now be seen in a range of new technologies from launch facilities to space tracking capabilities, communications infrastructure, and satellite production. One example was the YaoGan Weixing remote-sensing satellites, which China began to launch in April 2006. Its two variants—one with enhanced electro-optical features and another boasting China's first synthetic-aperture radar capabilities—now provided all-weather coverage and data precise enough to be used for certain military purposes.[129] These improved reconnaissance capabilities began to enhance China's defensive posture.

On the offense side, the Chinese military began a series of unannounced launches of a ground-based interceptor system into space in 2005 and 2006 to ready itself for the actual kinetic destruction of a satellite. These apparent "proximity tests" aimed at checking the system's radar seeker without actually colliding with a space object, which would have generated noticeable debris. Then, in January 2007, civilian space trackers in several countries noticed the absence of the Chinese *FY 1C* weather satellite, leading to Internet discussion.[130] This anomaly had followed the apparent launch of a two-stage DF-21 ballistic missile from a mobile launcher at one of China's known test sites. Finally, after strong speculations began appearing in various media outlets, the U.S. government released intelligence information about a kinetic ASAT test that China had conducted secretly some two weeks before. Beyond the secrecy of the test, this event damaged China's reputation in the United Nations as a promoter of space arms control, given the Chinese Foreign Ministry's initial public denial of the action, its subsequent silence for twelve days, and then its failure to release any data to the international community. The results of the test proved particularly harmful for the near-Earth orbital environment, creating some 2,700 pieces of trackable debris (larger than ten centimeters) and likely thousands of smaller pieces.[131] Debris experts called it the single worst kinetic event in space history, particularly considering that its altitude of some 525 miles meant that the debris would take more than forty years to deorbit.[132] Worldwide criticism of the test stung Chinese space officials, who seemed to have thought the world would accept the test, as it had two dozen Soviet tests from 1968 to 1982 and one U.S. ASAT test in 1985. But times had changed, and space was now crowded with many more high-value assets, including the permanently occupied *ISS*. China cancelled a UN debris working group meeting that had been scheduled to take place in Beijing to avoid further international criticism. When the head of U.S. Pacific Command, Admiral Timothy Keating, met his Chinese counterpart General Guo Boxiong in January 2008 and tried to discuss the ASAT issue, his questions were reportedly "dismissed with nervous laughter."[133]

The event appeared to show both the inordinate influence of the PLA on the space program—at least compared to the Ministry of Foreign Affairs— and China's failure to anticipate such negative foreign reactions. It also revealed a disconnect between the actual decision makers over space operations and China's State Council, whose White Paper "China's Space Activities in 2006" had pledged to "protect the space environment,"[134] although there was little doubt that President Hu Jintao had approved the test.[135]

Chinese analysts and officials attempted to place the blame on the United States for "forcing" it into the military space arena. They cited the numerous U.S. Air Force documents and the U.S. National Space Policy that discussed U.S. space "dominance" and policies of space "control." In addition, they pointed out that the United States has blocked progress at the CD in Geneva on a potential space arms control treaty. One Chinese space expert noted, "The United States has not been very responsive . . . for many years. I think China is justified in thinking that if the leading space power is unwilling to be limited then nothing is possible with arms control in space."[136]

Indeed, China continued to promote new forms of space arms control. In January 2008, China and Russia issued a revised proposal for a new arms control treaty offering clearer definitions and more specific clauses but still neglecting verification measures for the kind of intrusive knowledge needed to prevent the launch of a space weapon.[137] Chinese and Russian interlocutors, however, pointed out that the Outer Space Treaty also lacked a specific verification protocol. The treaty defined "use of force" and "threat of force" to mean "any hostile actions against outer space objects including . . . those aimed at their destruction, damage, temporarily or permanently injuring normal functioning, deliberate alteration of the parameters of their orbit, or the threat of these actions."[138] The Bush administration remained unmoved, although the State Department did respond in August 2008 with a six-page critique of issues the United States found objectionable, poorly defined, or simply confusing in the Russo-Chinese draft.

With the Obama administration in office and with a CD agenda agreed to in May 2009, Russia and China then issued a six-page letter clarifying what the treaty would and would not include in August 2009. To the disappointment of many in the international arms control community, the Sino-Russian response stated that the "PPWT does not prohibit the development of ground-based, water-based or air-based anti-satellite weapons systems because there is no way that such activity can be effectively verified. Nor does it prohibit the testing of such weapons by a State."[139] In essence, China's position had changed markedly since it had proposed its suggested elements for a new space treaty in 2001. China was no longer interested in the general nonweaponization of space but instead in a much more restricted ban on space-*based* weapons only, allowing it to keep developing its ground-based ASAT system. The draft treaty also, arguably, sanctioned the violation of the UN Debris Mitigation Guidelines of December 2007, by allowing kinetic tests against space objects that could create long-lasting debris. With these actions, China gave up the moral high ground it had once appeared to occupy regarding space arms control.

An Obama administration effort to engage the People's Liberation Army in space security talks moved forward in January 2011, when Secretary of Defense Robert Gates traveled to Beijing in hopes of starting a regular strategic dialogue that would include space. His Chinese counterpart, General Liang Guanglie, however, rejected a timetable for such meetings, citing disagreements over U.S. arms sales to Taiwan.[140] Nevertheless, the Chinese military agreed to a more limited set of future talks.

Commercial Space and Industry

In terms of industrial priorities, China's 2006 White Paper on space listed the aim to "build up the comprehensive national strength" as one of the country's core rationales for space activity.[141] More specifically, it noted, "China considers the development of its space industry as a strategic way to enhance its economic, scientific, technological and national defense strength."[142] Accordingly, China was moving by 2010 from being a relatively minor player in the commercial space field into a more central role. While it still lagged considerably behind major space-launching countries in terms of its number of commercial launches (with only occasional foreign payloads), it was catching up again in long-term orders, thanks to proven reliability and lower costs. Part of this shift resulted from a strategic move by major European satellite firms to develop technological independence in order not to have their launches limited by U.S. export control rules. The decision by the private French company Eutelsat Communications Group in 2008 to purchase insurance for up to nine satellites for future "ITAR-free" launches on Chinese rockets caused a political furor in the United States, as it represented the first time that a traditional U.S. ally had bucked the U.S. embargo on Chinese launches.[143] France's Arianespace also opposed the move, given its likely effect on its own launch business.[144] In 2009, Eutelsat specified that its first booking with GWIC under the plan would involve the Chinese launch of a French Thales-built satellite,[145] leading to a call by U.S. Congressman Dana Rohrabacher, a Republican from California, for economic sanctions on France.[146] But the sanctions did not move forward. Instead, two other major international satellite operators (Intelsat and SES) began to lobby the U.S. government for changes in U.S. regulations to allow contracts with both Chinese and Indian launch providers.[147]

Overall, Chinese-European space contacts had grown steadily since the late 1980s and now had formed a link that the United States had trouble breaking. In 2001, CNSA and the European Space Agency (ESA) signed an agreement to conduct cooperative research on the Sun's effects on the Earth's environment, in the so-called Double Star program.[148] In 2003 and 2004, China contributed satellites to this project's in-orbit monitoring network, marking one of the first significant operational missions CNSA has conducted with a major foreign space agency. These and other contacts influenced China's decision to base its mobile phone network on European technology rather than on the U.S. standard.[149] The one setback, ironically, took place because of funding problems within Europe for its planned Galileo navigation system. Initially, China pledged some $300 million in investment funds in order to become a full partner in the Galileo project, which Beijing viewed as a counterbalance to the U.S.-controlled Global Positioning System (GPS). But when adequate private financing could not be found in Europe and governments had to step in with funding to assure the system's development, European governments opted to place national security restrictions on the system. The Europeans eventually ousted China out of security concerns and irritation with Beijing's new plans to build a competing commercial system as part of its Beidou program.[150]

CURRENT POLITICS OF THE SPACE PROGRAM

Despite the emerging foreign image of China as a modernizing military juggernaut that uses its authoritarian state to channel resources in an efficient manner into priority areas such as space, Chinese authorities today see a different picture of themselves. They see a country beset by problems of political legitimacy (witnessed by the sharp rise in mass protests), Communist Party political decay (seen in the declining relevance of Marxist concepts to China's capitalist reality), rising state-level corruption (witnessed by widespread news on the theft of public funds by officials),[151] and worsening organized crime (evidenced by recent crackdowns in a number of cities). China also has a growing and possibly volatile migrant population, particularly if widespread unemployment emerges in any future economic downturn. None of these problems means that China cannot still devote significant national resources to space activity, but they do signify problems that may detract from space funding and make success more difficult to accomplish than outsiders might appreciate. As the former U.S. deputy assistant secretary of state for East Asia Susan Shirk writes: "Dogged by the specters of Mao Zedong and Deng Xiaoping, the revered leaders who preceded them, China's current leaders feel like midgets, struggling to stay on top of a society roiled by economic change."[152] Given this context, what are China's emerging priorities for space, and what problems is Beijing likely to face?

Commercial Space Priorities

The tremendous growth of new businesses using precision navigation that emerged in the United States and around the world after the U.S. completion of its GPS network has caused China to commit itself to building a commercial network called Beidou (Compass). It began launching the satellites—based on the *DFH-3* design—in 2000 and continues to populate the system, which is scheduled to become operational on a regional basis in 2012.[153] By mid-2011, it had launched its first eight satellites. When completed, the system will be configured with five satellites in geostationary orbit and thirty satellites in medium-Earth orbit,[154] where U.S. GPS satellites are located. Beidou is scheduled to offer global coverage by 2020 and will be in a favorable position to compete with smaller national systems planned by Japan and India. The constellation will also provide military benefits, giving the PLA far greater accuracy with its missiles. China has agreed to make its civilian system interoperable with GPS and other national systems, although one Japanese official has stated that "few, if any" talks have taken place yet with Japan.[155]

In terms of its space-launch capabilities, China had conducted over 135 successful launches as of early 2011. While Russia led the world with thirty successful launches in 2010, China's yearly rate had accelerated rapidly and for the first time equaled the U.S. rate, at fifteen launches.[156] Questions of human safety that have been raised in the past about Xichang and other launch sites may be mitigated by China's shift of major space operations to its new Hainan Island site in the coming years, which will launch over water. This facility will also allow China to launch payloads of larger size, as the site will be able to receive spacecraft and other cargo by ship, rather than facing current limits imposed by rail and tunnel transport from the production facilities in Beijing.[157] China aims to use the Hainan launch site to operate its planned larger rocket, the Long March 5, which has been delayed and is now scheduled to begin regular flights in 2014.[158] The new rocket, being constructed at CALT's space industrial facility in Tianjin,[159] will more than double existing payload-to-orbit capabilities by offering a capacity of twenty-five tons. It will also serve as the launcher for a planned space station.

China's rapid rise in space does not guarantee its success, and it remains behind world standards in a number of critical space technologies. Despite

Thailand's membership in APSCO, for example, Bangkok turned to a European consortium to purchase its *Thailand Earth Observation Satellite* (*THEOS*), whose remote-sensing technology was more sophisticated than China's.[160] As Beijing's 2006 White Paper admitted after noting its technological successes: "The development of the space industry in China now faces new opportunities and higher requirements."[161]

One problem that China may face in the future relates to its state-run model of organization. The steady privatization of a number of space activities in the United States raises a question: which model promotes space development better, socialist or capitalist? China's 2003 White Paper on space explained China's approach: "With the establishment and improvement of China's socialist market economic mechanism, the state guides the development of space activities through macro-control, makes overall plans for the development of space technology, space application[s] and space science, [and] promotes the R&D and system integration of important space technologies."[162] Whether the Chinese state is flexible and entrepreneurial enough to thrive in the next stage of international space competition remains to be seen. As Handberg and Li summarize: "The key will be whether the Chinese private space sector is able to free itself from government control and grow economically."[163]

Chinese lawyers have been struggling to develop a body of commercial space law to catch up with the country's growing space activities, which currently stymies development of a real private sector. As the Chinese space lawyer Li Juqian explains the existing legal situation: "The provisions on space activities fall into neither the realm of laws nor that of administrative regulations, but belong to the rules formulated by the State Council's ministries, which are at the lowest level of the legislative hierarchy."[164] Another Chinese lawyer, Li Shouping, predicts that as China becomes increasingly engaged in international space activities, it will have to improve steadily its legal framework for space and pass legislation to bring itself into greater harmony with global norms and practices.[165]

But China's state-led space industries might eventually stumble in the face of foreign competition. A 2010 study by the China expert Eric Hagt called China's space industry "dispersed, bloated, and located in geographically isolated regions."[166] The sector has also had to deal with a series of reforms as Chinese authorities have sought to inject greater civilian management and innovation into hidebound defense industries, which in the mid-1990s were "in a perilous state."[167] With this in mind, the Chinese State Council demoted the old umbrella organization for scientific research and development for the defense industry, COSTIND, in 2008. In its place, a new department called the State Administration for Science, Technology, and Industry for National Defense (SASTIND) has been created, with re- duced responsibilities and a subordinated role within the new super-Min- istry of Industry and Information Technology.[168] One of the survivors of these reforms, however, has been CASC, which now subsumes the launch industry actors GWIC and CALT, as well as nine academies, a half-dozen enterprise groups, and some three hundred enterprises throughout the country, with approximately 100,000 total employees (not all of whom work in the space field).[169] While CASC lost some power after a 1999 reor- ganization, it retains a broad area of responsibility that includes "research, design, manufacture and launch of space systems such as launch vehicles, satellites and manned spaceships as well as strategic and tactical mis- siles."[170] CASC also retains responsibility for the Shenzhou manned-space- flight system. The other major actor is the China Aerospace Science and Industry Corporation (CASIC), formed in 1999. According to a study by Mulvenon and Tyroler-Cooper, CASIC shares an emphasis on missiles, al- though "its business areas also include satellite R&D and delivery systems, and military and civilian applications of information technology."[171]

Yet while China's space enterprises are seeking profits abroad, politics matter too. As the China analyst Kevin Pollpeter argues, "Commercial space services not only increase revenues for the space industry but have also been used to advance China's diplomatic interests with oil-rich coun- tries"[172] such as Nigeria and Venezuela. In both cases, these deals for Chi- nese-built and launched geostationary communications satellites were formally commercial but on very favorable credit terms to the purchasing countries, with China providing some costs and offering low- or zero- interest rates on its loans. The total value of the *Nigcomsat-1* (launched in 2007) deal was $311 million,[173] and the contract for *Venesat-1* (launched in 2008) was $241 million.[174] China offered technical training to each country's space scientists and built ground stations on their territories. In a number of deals, launch costs have also been covered by China. This strat- egy offers political benefits but will impose costs on the Chinese govern- ment and space industry. The *Nigcomsat-1* had a faulty solar array, and the spacecraft ceased functioning in 2008. Beijing has had to offer a replacement satellite. In other commercial deals, China has contracted with Pakistan to build and launch the *Paksat-1R* geostationary communications satellite.[175] It has reached similar deals with Laos to build and launch *Laosat-1*[176] and with Bolivia for the *Tupac Katari* communications satellite.[177]

Another example of China's use of space to promote its political interests is the country's history since the mid-1990s of contracting with Intelsat to make direct-to-home broadcasts of China Central Television available in the United States and approximately one hundred other countries.[178] While not dissimilar from service provided by the BBC and Voice of America around the world, this contract—and its recent upgrading to offer twenty-four-hour programming—shows that China sees space as a key medium for making sure that its own political interpretation of events is available (and propagated) abroad.[179]

Although China and the United States have argued over Beijing's past missile sales abroad, the State Council and the Military Commission of the Communist Party's Central Committee have since 1997 implemented new export controls and a licensing system. Since 2002, the Military Products Export Control List—administered by SASTIND—has included a special Category 8 for military space items. According to Li Shouping, "In this category, strict regulation has been made on the export of military satellites and carrier rockets, which also provides a reference for the management of related civilian space products."[180] While likely to meet with some criticism from foreign observers, this recent development of space-related export controls must also be viewed as a positive effort to come into greater compliance with international standards and efforts to prevent the proliferation of space products that could be used for military purposes, such as delivering weapons of mass destruction.[181]

Civil Space Plans

China's priorities for the coming five years in space applications, as announced in 2008 by then-CNSA Administrator Sun Laiyan, include the development of higher-resolution remote-sensing satellites and related ground stations as well as spreading the benefits of China's Beidou navigational system by putting it into regionwide operation.[182] In space exploration, China is cooperating with Russia on a mission to the Martian moon Phobos, and it launched a second lunar orbiter (*Chang'e 2*) in October 2010 as a preparatory step for a lunar-lander mission and a later sample-return mission in 2017 to 2020.[183] China also is working to develop and deploy a series of three small space laboratories (*Tiangong-1, -2, and -3*) in the coming decade, with the first two due to be launched by 2013.[184] Further ahead, China has announced plans for a thirty-ton space station to be launched by 2022.[185] Some officials have mentioned a possible 2024 Moon mission as well.[186]

An important question remains as to whether meaningful civil space cooperation and progress is possible between CNSA and NASA, despite the optimistic statement on space at the 2009 Beijing summit and NASA Administrator Bolden's trip in 2010. One Chinese space scientist, Yi Zhou, argues that the problem relates to an organizational mismatch in current efforts, observing: "China's civil space program is organized differently from that of the USA. Although CNSA is the Chinese national space agency, it does not have its own research institutes, as does NASA."[187] For this reason, while Zhou predicts that "the future is promising" for Sino-U.S. space ties, he argues, "There is no obvious way to jump-start actual cooperation in a short period of time."[188]

MILITARY AFFAIRS AND SPACE

While it is considered normal for large powers today to have military capabilities in space, foreign observers of China's military space program are engaged in a debate as to whether the country is also committed to building up full-scale capabilities for space warfare or instead seeks a more moderate "force support" capability as a "hedge," with a limited ASAT program. Supporters of the first interpretation point to traditional Chinese opacity regarding their nuclear and other strategic military assets and the Chinese military's active discussion of space warfare scenarios.

On the other side of the argument, Handberg and Li make the case that "So far China has made space observation, navigation/positioning, and communications the first priority."[189] These are normal military support functions found in many space programs around the world. While some media reports have written about China's alleged use of a laser to "blind" a U.S. satellite, the then-head of U.S. Strategic Command General (USAF) James Cartwright failed to confirm these charges in a 2006 interview, noting instead merely the existence of such Chinese research facilities.[190] Thus, while China continues to develop its military space capabilities, it has not used these technologies aggressively so far, nor has it pursued a strategy of concentrating its assets in military programs so as to achieve (or attempt to achieve) dominance in space.[191] Areas of continued Chinese military space weakness pointed to by foreign analysts include the absence of signals or electronic intelligence satellites; the lack of a space-based missile early-warning network; the lack of an integrated, tested, and secure military-use GPS system for precision missile guidance; a limited constellation of milsats (heightening vulnerability to attack); limited military experience in modern space operations; apparent possession of inadequate number of launchers for a large-scale ASAT attack; and the lack of a major military ally in case of extended space warfare.[192] Despite frequent complaints about China's rising capabilities, no U.S. military leaders would trade places with their Chinese counterparts. Thus, questions remain about China's commitment to a full-scale space arms race, and it is likely that both domestic economic factors as well as international circumstances will influence China's future direction.

Looking ahead, if China is seeking to "leapfrog" in terms of military space, one indicator may be its progress in developing microsatellites, which could offer a cheaper way to build capability. This will involve developing innovative technologies, testing these new systems, and deploying them in significant (and dispersed) numbers. As James Lewis observes, "Test deployments by China of microsatellite systems for military communications, reconnaissance, or SIGINT [signals intelligence] would indicate a decision to seriously pursue this approach to space."[193] To date, there is some evidence of microsat activity (as in all advanced countries) and recent success in using the *BX-1* for space-based observation. In the summer of 2010, China also tested the ability of two satellites to engage in a rendezvous in space, with possible civilian and military purposes. But this falls well short of a full-blown offensive program, and not all technologies work as well when miniaturized (highly accurate reconnaissance and radar satellites being two examples).

How then should we best characterize China's military space capabilities and intentions? As noted, China's technology is still weak in a number of areas of military space activity. There is also little hard evidence that some of the PLA's suspected military space programs exist or a clear operational sense of how China would conduct any sustained military campaign against U.S. and other foreign assets in a noncooperative context, such as would certainly exist in a wartime scenario. But China's defense budget continues to expand in tandem with its economy, growing by low double-digit percentages each year. Still, Chinese writings on the subject of space warfare predominately lie in the conceptual realm and focus almost exclusively on planned or projected U.S. space capabilities and thinking in U.S. publications, in which there is ample evidence of detailed discussion of space weapons. As Zhang and McClung observe regarding Chinese writings: "examination of the sources indicates that China continues to borrow heavily from the language and the rhetoric of US concepts while continuing to struggle in the development of its own theories and strategies for space warfare and counterspace operations, lagging even further behind in its ability implement them."[194] Wortzel also observes: "the PLA's terminology often flows from what its officers read in U.S. doctrine."[195]

In the end, Chinese decisions to develop and deploy military space capabilities will involve strategic choices. Handberg and Li conclude: "With states such as India (a regional rival) openly espousing the concept of military space operations, the pressures will mount in support of hiking the military side of the equation."[196] But responding with a full-scale offensive program will have costs, both in terms of China's effort to present itself as a friend of peace regionally and internationally and in the form of high expenses for these technologies, which may or may not ever be used. In addition, Chinese decision makers will face difficult tradeoffs regarding heightened military space activity, if such a course means they have to cut back on higher-prestige civil space programs like lunar exploration or human spaceflight.[197] China's answer will therefore be affected by future assessments of the threat environment, prospects for managing China's space security concerns through diplomatic means, and the relative priority of space cooperation compared to competition.

China's restraint from testing its kinetic ASAT system against additional orbiting satellites—which generated broad international criticism in 2007 because of the debris that the test generated—suggests that Chinese space decision makers may be learning and altering their behavior over time. Some speculation after a Chinese missile-defense test in January 2010 suggests that the PLA may be moving its ASAT tests instead to lower altitudes and engaging nonorbital targets to reduce their debris profile. Such practices are more consistent with long-established U.S. missile-defense tests. Whether positive foreign engagement and a less hostile international environment might help promote further restraint in Chinese space behavior is an open question.

CONCLUSION

After two decades of halting progress in space activity until the mid-1970s, Deng Xiaoping's economic ideas and experiments with worker incentives proved successful not only in propelling China into rapid development over the next three decades but in restarting its space program: first in commercial applications, then in wide-scale civil space activity, and later in military programs. China's space program began to be noticed in the 1990s for its launch systems and, during the years since 2000, for its satellites, major accomplishments in human spaceflight, and limited (but threatening) offensive military technologies.

A significant advantage of China's space activities is its role in fulfilling domestic economic and political goals in the context of a rapidly modernizing society and the dwindling relevance of communist ideology for a rising and educated middle class. As Pollpeter notes: "Domestically, by developing a robust space program and participating in high-profile activities such as human space flight, the Communist Party demonstrates that it is the best provider of material benefits to the Chinese people and the best organization to propel China to its rightful place in world affairs."[198] But Wortzel sees the hostile Asian region as a factor in motivating China's military space effort, noting: "China's neighbors are [also] developing the sort of space warfare capabilities that the U.S. and the Soviet Union considered decades ago."[199] In addition, unlike its regional rivals, India and Japan, China has to be concerned about maintaining its nuclear deterrent against the United States, as well as preparing for a possible Sino-U.S. conflict over Taiwan. Thus, China is playing a more complicated two-level game in regard to space, and it is playing its side without major allies.

Perhaps for these reasons, China has long expressed an interest in building a stronger treaty-based security structure for space. To date, it has pursued the PPWT proposal. But while China initially won some sympathy from countries that saw the Bush administration as having aggressive intentions in space, its own 2007 ASAT test and 2009 paper offering clarifications about the PPWT seem to have squandered this initiative. How China proceeds in the future remains to be seen. The hesitancy of the PLA to engage in military-to-military discussions on space security remains a troubling precedent. But if China's secrecy stems from a lack of preparation or uncertainty, there is a prospect that the gradual expansion of its capabilities and its increasing dependency on space may lead to new incentives to participate in space tension reduction. In the area of export controls, for example, the China analyst Stephanie Lieggi has noted a similar disinclination by Beijing to engage in broader regional cooperation and discussions. But, she argues that "this hesitancy may change as China's own [export control] system continues to improve and Chinese officials recognize the value of having trading partners with stronger trade control systems. At that point, Beijing may see a definite benefit in being a 'model' citizen in the global nonproliferation regime."[200] If, however, China's disinclination to engage in these discussions is rooted in a fundamental rejection by the PLA of negotiations with the U.S. and other militaries, then the future of space security is likely to be far more bleak, and the prospects for avoiding a Sino-U.S., Sino-Indian, and even Sino-Japanese arms race in space are unlikely.

These points again highlight the importance of politics and economics as the other two sides of the triangle of strategic interests, besides military affairs. If political relations remain generally positive and if economic relationships within the region and with the United States remain productive and mutually beneficial, the prospects for military rapprochement will improve. But if unresolved regional tensions remain a festering part of the military equation, then China may continue to treat space as an area where compromises are not possible. This dynamic highlights the importance of addressing underlying regional factors in space and the need for new management mechanisms between China and its regional rivals, including the United States. These topics constitute the central focus of the final chapter.

CHAPTER FOUR

THE INDIAN SPACE PROGRAM

Rising to a Challenge

India's place in the world of major space powers is unique. It has been a nuclear-armed country since 1974, but its space program has had a remarkably peaceful orientation throughout most of its history. In contrast to the Soviet Union, the United States, and China, its fleet of space-launch vehicles originated as civilian rockets, not as military-purpose ballistic missiles converted to space-launch use. Moreover, until very recently India lacked any appreciable military space program, having focused for decades on Earth remote sensing, communications, and weather forecasting to serve the civilian economy and provide benefits to India's vast and dispersed population.

However, China's rise, the U.S. rapprochement with India, and India's general effort to increase its indigenous military technology have put pressure on its space program's traditional profile. In recent years, India has begun to take military uses of space increasingly into mind as it develops capabilities for the twenty-first century. It has launched advanced reconnaissance satellites (drawing on its partnership with Israel), moved forward toward deploying a national system for precision navigation, opened a military space command, and spoken openly of future kinetic antisatellite capabilities and possible laser weapons.[1]

China's efforts in space exploration—particularly its human spaceflight "firsts" for Asia—have also put India on notice that it risks falling far behind if it does not move quickly to rectify the situation with its own programs. Accordingly, India has pushed forward in recent years with a major expansion of its civil space program, particularly in high-prestige lunar-oriented research. Its 2008 *Chandaryaan-1* mission represented a major breakthrough for India in orbiting its first lunar satellite, which hosted over a dozen experiments (including two for NASA). A major debate in India is ongoing regarding parameters of a human spaceflight effort. These factors suggest that India has recognized that its traditional focus on space applications will not be adequate if it seeks to compete with its regional rivals, promote its national security, and establish a respected reputation internationally for its space progress.

Finally, India's advancing capabilities in the information technology sector have helped push a realization that India's commercial future must include a role for the space sector as well. For these reasons, India has now become increasingly active in promoting its space-launch vehicles on the commercial market and is beginning to offer space services in the satellite field as well. To date, however, U.S. export controls have prevented many countries that use U.S. technology from launching on its rockets.

This chapter analyzes India's space program from the context of its domestic, regional, and international aspects. India's status as a postcolonial, nonaligned, and developing country affected its initial direction in space activity, and its strategic rivalries with Pakistan and China have recently caused it to move toward a more active military stance. These strategic competitions represent one of the most risk-prone areas of Asia's space competition. At the international level, the Indian relationship with the United States has sought to balance the rise of China. Their common concern and the two sides' partnership in counterterrorism efforts have pushed New Delhi and Washington into a return to the relatively close relationship that prevailed between the Indian Space Research Organisation and NASA in the 1960s and 1970s, although India now has a broader base of independent capability. India still lags behind China in a variety of space technologies but is competitive in some areas. The difficulty, as has been the case historically, has been maintaining the firm commitment and funding required to see programs through to their fruition. But India's dynamic information technology sector, its slowly internationalizing educational system, its rising middle class, and its increased recognition of its military's needs for first-class support technology from space suggest that the past may not be a mirror of the future. At the same time, India's fierce independence, tendency for hiding mistakes, and gap in capabilities compared to such leading space powers as the United States, ESA, and Russia mean that it still faces significant challenges as it seeks to establish itself as a more accomplished space actor across the range of space sectors.

INDIA'S EARLY SPACE EFFORTS

India's space program has deep roots, although the pace of its progress has been hampered by India's poverty, initial gaps in its technological infrastructure, the complexities of Indian domestic politics, and strategic factors within South Asia. Nevertheless, India historically has been one of Asia's leaders in the field, particularly in space applications.

The origins of Indian interest in space and rocketry can be traced to at least two sources. In terms of astronomy, Indian scientists (such as Aryabhata) established a tradition of space-related inquiry as early as the 400s C.E.,[2] and Indian observatories founded at Jaipur, Delhi, Mathura, Ujjain, and Benares in the early 1700s under Raja Jal Singh[3] and in the 1800s near Bombay and in Kodaikanal offered a high state of professional expertise that impressed European visitors.[4] In the military field, Indian forces under Hayder Ali and then his son used rockets constructed out of bamboo and iron—with a range of over one kilometer—against British forces in the late 1700s, eventually boasting a "corps of rocketeers" with five thousand men.[5] As Indian Air Force Wing Commander K. K. Nair writes: "While the Chinese pioneered the early development of rocketry, it was the Indians who first inculcated this revolutionary technology of rocketry into their military doctrines and regularly fired them in battles."[6] But India's prowess in rocketry was not enough to prevent its colonial subjugation in the mid-1800s, as the more rapid progress of British military technology overall made London's eventual victory all but inevitable.

In the area of science, British rule led to clashes between European and indigenous traditions, with the former eventually prevailing in almost all areas, particularly those that involved technology. India's elite youth also began to seek education in England as a means of advancement. Decentralization of many governmental functions as part of post–World War I reforms adopted in 1919 damaged Indian science significantly, as central support had always been critical for research.[7] These new circumstances increased the dependence of Indian scientists on British sources of funding, but only a handful of top researchers could hope to compete effectively for support.

The Nobel Prize awarded in 1930 to C. V. Raman helped restore the pride of Indian science, although economic hardships and World War II intervened to prevent any significant positive effects from taking root.[8] Indian scientists were not enlisted to contribute to Allied technological efforts during World War II.[9] The father of India's space program, Vikram Sarabhai, had studied at Cambridge in the late 1930s and worked in C. V. Raman's laboratory during the war.[10] After the end of hostilities with Japan, Sarabhai returned to Cambridge to complete his doctorate in physics, then established the Physical Research Laboratory in India to investigate atmospheric physics and cosmic rays.[11]

Only in the immediate aftermath of the war did Indian science see an upsurge, although its main trends reflected not so much Mahatma Gandhi's vision of a peasant society but instead the ethos of European positivism via technological advance. The historian David Arnold writes that Indian scientists in the 1940s "were disposed to believe that science and technology, rightly applied, could rapidly improve India's fortunes and remove many of its longstanding problems."[12] India's independence in 1947 saw these concepts become state policy, and Jawaharlal Nehru's emergence as the country's first prime minister brought a more technocentric leadership into power.[13] As Nehru stated: "The future belongs to science and those who make friends with science."[14] Gandhi's assassination in early 1948 decreased the peasant focus of the Congress party and increased the power of its secular and progressive elements, whose policies emphasized large-scale, state-run development projects.[15] Given the magnitude of tasks that faced the Indian government, space research remained in the background during the late 1940s and 1950s. A vast gulf remained between the dreams of enthusiasts and any practical chance of developing such sophisticated technologies as space rockets and satellites.

However, *Sputnik*'s launch in 1957 inspired scientists and technicians to form the Indian Astronautical Society in 1957 to stimulate domestic research on space topics and to help unify research efforts being conducted by various groups around the country.[16] In 1962, the Indian government formalized its participation in space research by establishing the

Indian National Committee for Space Research, beginning with work on sounding rockets at the Thumba Equatorial Rocket Launching Station (TERLS).[17] Sarabhai outlined the priorities for India's space efforts in 1962 by rejecting any notion of competing with the "economically advanced nations" in any sort of space race. He argued instead that India "must be second to none in the application of advanced technologies for the real problems of man and society."[18] This theme would guide India's space efforts for the next forty years.

In the same year, the socialist-leaning Nehru government signed a cooperative agreement with the Soviet Union to allow the Soviets to launch scientific sounding rockets from the TERLS site, located near the equator. India began the slow process of gaining expertise toward the eventual development of an indigenous space-launch vehicle, although the process would take nearly twenty years. The TERLS rockets were solid fueled and performed astronomical experiments while also studying the upper atmosphere and ionosphere; they carried payloads of up to four hundred kilograms.[19] As a leading member of the nonaligned movement, India sought a middle ground between the two superpowers to avoid a dependency position. During the early 1960s, Indian space scientists went to NASA's launch facility on Wallops Island, Virginia, for training in sounding rocket research. India also accepted NASA assistance in the form of Nike-Apache rockets.[20] India's first "modern rocket" to be launched from the Thumba site in 1963 was one of these rockets.[21]

One of the reasons for India's willingness to put off development of its own long-range missiles and to focus on space applications from rockets was its lack of nuclear weapons and its rejection of the superpower arms race. By choosing his policy of nonalignment, Nehru also sought to avoid being drawn into a possible third world war.[22] Thus, unlike in the Soviet Union, the United States, and China, national space priorities did not initially hinge around the development of an intercontinental delivery system. Moreover, India's space leaders rejected the notion that a developing country could afford to have a prestige-oriented program.[23] The Indian space engineer G. B. Pant critiqued the high costs of the U.S. lunar program at India's first national symposium on rocketry in 1967: "Is this a valid enterprise? Could not this effort be applied for the teaming, starved, illiterate, ill housed, ill clad, ill cared [for] population of the world?"[24] Instead, Sarabhai and his colleagues decided to focus on technologies that could provide functional benefits for India's economic development, educational system, and health care provision. As a former Indian space official recounted: "Due to their enormous potential for enhancing the quality of people's lives, three sectors were hand-picked for development: satellite communications, weather/climatology, and remote sensing."[25]

India's space program had been placed administratively under the similarly development-oriented Department of Atomic Power, with Sarabhai as the head of space activities. In order to tap into existing space assets, Sarabhai began to build ground stations to receive signals (including television broadcasts and weather data) from U.S. and Soviet communications satellites and the Intelsat network, with equipment initially obtained from Japan's Nippon Electric.[26] India soon developed its own ground-station technology and began to expand its network in order to spread the benefits nationally from the communications revolution that had already begun in the developed world.

In 1969, the government separated space from the nuclear energy program and established the Indian Space Research Organisation (ISRO) to serve as the primary functional body for space technology development and applications, with Sarabhai becoming its first chairman. ISRO was tasked with the eventual development of a satellite launch vehicle to allow India to achieve independent access to space. After India's victory in the 1971 war with Pakistan over the founding of an independent Bangladesh, the Indira Gandhi leadership formed a supervisory body in 1972 called the Space Commission and an administrative body called the Department of Space, thus removing ISRO from the Atomic Energy Commission and linking it more directly to higher levels of the government.[27] This new attention helped initiate a 1973 strategic plan for ISRO to work toward launching a forty-kilogram satellite into low-Earth orbit.[28] In the meantime, ISRO continued to expand its network of ground stations and began to work toward the construction of indigenous satellites for initial launch by other countries. It also increased staffing for the space program to ten thousand persons by the mid-1970s, including those based in ISRO, universities, and industry.[29]

Sarabhai recognized the difference between India's needs as a developing country and the traditional path followed into space by prior space powers. As the Indian space analyst V. S. Mani argues, Sarabhai "favoured 'leap-frogging' the process of development for developing countries by acquiring and developing competence in advanced technology for the solution of their particular problems"[30] rather than focusing on space science and exploration, which were seen as superfluous pursuits. For India, this realization led directly to Sarabhai's backing for the Satellite Instructional Television Experiment (SITE) with NASA in 1975 and, later, the promotion of the use of foreign Earth remote-sensing data to aid in domestic crop estimates, disease monitoring, disaster relief, location of fishing stocks, mapping, and tracking of natural resources such as forests.[31] The SITE program allowed regional broadcasts of instructional materials in a variety of educational fields of relevance to the economy. During 1975 and 1976, ISRO conducted a major experiment in six Indian states, with 2,400 villages that had been equipped with direct reception sets.[32] This trial proved the effectiveness of using satellites to deliver educational programming to rural areas for the improvement of "health, hygiene, family planning and better agriculture practices."[33] The SITE project later became a model for the wider delivery of government-sponsored educational broadcasts throughout the country.

In remote sensing, India benefited from an agreement reached with the United States, which began to offer Landsat data on the Indian subcontinent via a complete U.S.-provided ground station that opened in Hyderabad in 1979.[34] Although the program provided highly beneficial data, coverage was sporadic, given the nature of Landsat's orbital coverage of India. But the facility also benefited from data provided by the U.S. National Oceanic and Atmospheric Administration (NOAA).[35]

India's foreign relations and its unique position between the two superpowers assisted its efforts to acquire space technology and know-how. Indian scientists benefited from contacts with NASA scientists in developing its system of ground stations and later worked with General Electric, Hughes Aircraft, and MIT experts to gain knowledge for the design of the first-generation Indian satellite (Insat) system.[36] The United States continued to supply equipment and components for India's communications and remote-sensing satellites even after the 1974 Indian nuclear test. In 1972, India also reached a highly favorable agreement with the Soviet Union for access to Soviet launch vehicles for its satellites, in return for Soviet access to Indian ports by Soviet space-tracking ships. As Sundara Vadlamudi writes of this period: "During India's formative phase, its space program benefited greatly from foreign assistance, chiefly provided by the United States, the USSR, Japan, West Germany, France, and other countries."[37]

With the offer of a Soviet launcher as a carrot, Indian scientists raced ahead with the development of their first indigenous satellite, a research spacecraft named *Aryabhata* (after the Indian astronomer).[38] The satellite entered orbit aboard a Soviet Cosmos 3M rocket in April 1975, but its solar-electric system failed after six orbits. ISRO scientists learned from this mistake in the redesign of their second satellite, *Bhaskhara-1*, for which they acquired more sophisticated instrumentation, including an electrical system that benefited from 3,500 Soviet solar cells.[39] The Soviet Union launched *Bhaskhara-1* in 1979. While the new satellite experienced a few technical problems that limited its functionality, it provided one-kilometer-resolution television images of the Indian landmass and beamed meteorological data to seven weather stations around the country.[40] A follow-up satellite, *Bhaskhara-2*, provided similar services, along with enhanced mapping capabilities for determining rainfall rates and water vapor.[41]

Indian launch-vehicle development had begun to make progress by the late 1970s, with tests of the four-stage, solid-fuel Space Launch Vehicle (SLV)-3 beginning in 1979 at the new Indian launch site on the barrier island Sriharikota, located east of Bangalore (between Pulicat Lake and the Bay of Bengal). Based on the U.S. Scout missile, the SLV-3 had four solid-fuel stages, and 85 percent of its components were domestically manufactured.[42] The rocket's first attempted launch with a satellite aboard in August 1979 failed five minutes into the flight, and it fell into the Bay of Bengal. The SLV-3's second attempt in July 1980 succeeded in orbiting the small, thirty-five-kilogram *Rohini-1* satellite, which entered into an elliptical orbit and communicated data to Indian ground stations.[43] ISRO followed with the *Rohini-2* satellite in May 1981, which carried a small solid-state imaging system.[44] Although the SLV-3's payload capacity limited the applications available to it, India had now graduated into the class of countries with independent access to space. To begin to build toward its next stage—a geostationary satellite—India received assistance from France in launching the *Ariane Passenger Payload Experiment* (or *Apple*) experimental communications satellite, which entered geostationary orbit in June 1981 along with France's *Meteosat-2*.[45] The *Apple* project provided India with experience in testing equipment and communications involved in operating a geostationary comsat.

Indian efforts to establish its capabilities for operating communications satellites also benefited in 1977 to 1979 from cooperation with the

Franco-German *Symphonie* satellite.[46] But in order to establish a devoted satellite communications network in orbit, India contracted with the U.S. Ford Space and Commerce Corporation for the first seven satellites of its planned Insat system,[47] although with the eventual aim of building its own satellites. NASA launched the first *Insat-1A* spacecraft in April 1982, which began to connect the thirty-one ground stations built to help link the country's communications to the space network.[48]

Although U.S.-Indian relations remained positive during the first three years of the Carter administration, several factors led to a serious decline in bilateral ties: the Soviet invasion of Afghanistan in December 1979, the U.S. decision to work with Pakistan to aid Afghan rebel forces, Indira Gandhi's return to power, and her administration's decision to side largely with Moscow over the conflict.[49] Meanwhile, relations with the Soviet Union improved, and New Delhi took up a Russian offer to train and fly India's first (and thus far only) astronaut aboard the *Salyut 7* space station.

The Indian astronaut selection process began in 1981, when ISRO whittled down a list of 150 applicants to eight and sent six the following year to the Soviet Union for initial training and further tests.[50] The Soviets narrowed the list to two air force pilots and began training them at the cosmonaut center Star City in the fall of 1982. Rakesh Sharma emerged as the top candidate for the flight and blasted off into space on April 3, 1984. Amid much fanfare in the Indian press, he completed a series of Earth-resources and medical experiments and returned to Soviet Kazakhstan successfully after a week in space. Despite political differences over Afghanistan, the U.S. Reagan administration extended an offer to India to fly a payload specialist on an upcoming U.S. space shuttle *Challenger* flight and to deploy India's *Insat-3* satellite.[51] But the explosion of the *Challenger* before the mission could take place and an Indian decision to carry out its own launch of the *Insat-3* cancelled this cooperative effort, leading to the two space programs' further estrangement.

MOVING TOWARD SELF-RELIANCE

In the mid-1980s, India began to seek more indigenous satellite capabilities and more powerful rockets.[52] It also faced new international obstacles, particularly in missile technology. As Vadlamudi recounts, "India's initial efforts to achieve self-sufficiency [had] focused primarily on rocket systems since the Indian scientists anticipated restrictions on such systems."[53] Unlike in most countries, India's technology transfer moved from the civilian side to the military side, rather than the reverse. As Dinshaw Mistry observes, "the DRDO [Defence Research Development Organisation]-based missile program borrowed human resources and technology from ISRO."[54] This included the wholesale transfer of the SLV-3 project leader, Abdul Kalam, to DRDO, whose work succeeded in 1989 in bringing about the flight of the first Agni-1 ballistic missile, which relied on the SLV-3's first stage.[55] Meanwhile, work continued at ISRO for a new space booster. Test flights of the Augmented Space Launch Vehicle (ASLV)—a five-stage, solid-fuel rocket based on the SLV-3 but with two strap-on boosters—began in March 1987.[56] The new rocket experienced two failures before its first orbital launch in May 1992, although the orbital insertion of its satellite payload was not fully successful. Finally, the ASLV achieved its goals in its last flight in April 1994. The ASLV could launch a payload of one hundred kilograms into a 450-kilometer orbit.[57] But India needed to be able to launch much larger satellites to improve their functionality and eventually allow it to place satellites in geostationary orbit, thus removing its dependency on foreign space programs. For additional power and control, ISRO finally turned to liquid-fuel engines for the second stage of its next launcher: the Polar Satellite Launch Vehicle (PSLV), which was capable of lifting a one-ton payload into orbit.[58] However, as India began to expand its missile program by drawing technology from its civilian rockets, the United States and other Western countries ceased technological cooperation with India for fear of promoting a nuclear weapons delivery system. The United States also began to back away from its previous willingness to provide India with advanced computers and other control equipment.[59]

Meanwhile, India had begun to meet its goals of developing indigenous satellite technology. In 1988, ISRO completed construction of its first *Indian Remote Sensing (IRS)-1* satellite, which the Soviet Union launched aboard a Vostok rocket. This large spacecraft, equipped with two deployable solar arrays, began to provide India with a regular and devoted data stream on Indian Earth and water resources via a sun-synchronous polar orbit.[60] India supplemented these systems by reaching agreements to receive all-weather synthetic-aperture radar information on ocean conditions, geology, moisture levels, and vegetation from Europe's *Earth Resources*

Satellite-1 (launched in 1992) and Canada's *Radarsat-1* (launched in 1995).[61] At the same time, India began to benefit from the gradual population of its geostationary Insat telecommunications constellation with a series of launches on French Ariane and U.S. Delta rockets beginning in 1988, which provided devoted streams of Indian television and radio programming as well as digital data transmissions throughout the country.

THE SOVIET COLLAPSE AND INDIA'S STRATEGIC REORIENTATION

The break-up of the Soviet Union in December 1991 created a dilemma for India's space program. Given Moscow's new poverty and requirement of hard-currency terms for arms and technology transfers, India now had even greater trouble acquiring critical hardware and experience in those areas of space activity where it still lacked domestic expertise. Under President Boris Yeltsin, Moscow sought financial assistance from the United States, causing it to hew to Washington's priorities, including those associated with the 1987 Missile Technology Control Regime (MTCR). Given the priority of nuclear and missile nonproliferation efforts under the Clinton administration, moreover, U.S.-Indian relations did not improve markedly either, despite Washington's cooling of relations with Pakistan.

At the same time, India's new Congress-led government under P. V. Narasimha Rao had entered office in July 1991 facing a welter of problems, particularly on the economic front. Like the centrally planned Soviet economy, India's state-led socialist model had ground the country to a virtual halt and frozen it out of major international lending organizations, making it a poor competitor for foreign trade and investment. In this context, India faced a bleak future if it continued on its existing path. Instead, Rao's government, led by Finance Minister (and later Prime Minister) Manmohan Singh, announced a series of major reforms, including devaluing the rupee, privatizing a number of state corporations, cutting state spending, and loosening controls on the private sector and foreign investment.[62] As Ranbir Vohra summarizes, these reforms showed "a deliberate decision to move away from the Gandhian belief in self-reliance, which had resulted in isolationism, and Nehruvian socialism, which has resulted in economic stagnation and low productivity."[63] The changes would set India on track for the information technology revolution that was to occur in the following two decades, the surge of investment into India, and the eventual rapprochement in relations with the United States. As for space, the thrust of these reforms placed a new emphasis on privatizing technologies developed by ISRO, leading the Department of Space to create the Antrix Corporation for this purpose in September 1992.[64]

In 1993, India's space-launch capabilities took a major step forward with its first successful flight test of the PSLV, although a software problem kept it from delivering its payload into orbit. Still, this 44.4-meter-tall, 294-ton rocket marked the maturation of India's space-launch program, with its solid-fuel first stage (consisting of a central core and six strap-on boosters), a liquid-fuel second stage, a solid-fuel third stage, and a liquid-fuel fourth stage. [65] Its second flight in 1994 succeeded in placing the *IRS-P3* satellite into a polar orbit,[66] and the launcher continued to build its record of reliability through the 1990s. Significantly, by 1999, the launcher offered India its first commercial successes in carrying South Korea's *Kitsat-3* and Germany's *Tubsat* into orbit.[67] A number of commercial flights followed for European, Latin American, and Southeast Asian customers. The PSLV also launched an Indian meteorological satellite in September 2002 and began to offer the ability to place satellites into a geostationary transfer orbit for insertion into the geosynchronous belt.[68]

A key element of India's effort to achieve a full spectrum of space capabilities was its desire to develop a launcher capable of placing payloads directly into geostationary orbit (22,300 miles above Earth), which would remove its dependence on foreign launch-service providers. After ISRO initiated negotiations with the Soviet commercial space agency Glavkosmos in 1988, the two sides announced an agreement in 1991 in which the Soviet Union would sell cryogenic liquid-fuel engines and associated production technology for the upper stage of India's planned Geosynchronous Satellite Launch Vehicle (GSLV).[69] Given India's use of the SLV-3's first stage in the new Agni ballistic missile—tested in 1989—the United States worried that the deal would provide India with technology that could be used to develop a long-range ballistic missile for nuclear weapons.[70] Following the Soviet break-up in December 1991, the George H. W. Bush administration sanctioned Glavkosmos in May 1992 over the India agreement, barring U.S. firms from cooperating with it. Given its desperate need for hard currency, the Yeltsin government initially rejected a U.S. request to cancel the Indian rocket deal. After steady American lobbying and the U.S. delivery of a substantial financial aid package to Russia at the Vancouver Clinton-Yeltsin summit in April 1993, the Russian government eventually agreed to amend the Indian deal by withholding the cryogenic production technology and selling only the completed boosters themselves, although some experts believe that the technology had already been transferred.[71] India's first set of flights with the GSLV, beginning in 2001, used Russian upper-stage engines,[72] although ISRO remained resolved to develop its own cryogenic engine technology.

Elsewhere, India reached out successfully under Prime Minister Rao in 1992 to establish mutually beneficial diplomatic and strategic ties with Israel,[73] another pariah from the nonproliferation regime. The two sides began to cooperate in weapons development and related trade, which would eventually lead to Indian access to Israeli expertise in space technologies as well. Israel's satellite reconnaissance technology became an object of particular interest, given India's relatively poor resolution in its existing Earth observation satellites.

Despite a series of broad bilateral agreements on technology transfer with the United States during the Reagan administration,[74] U.S. export control officials now approved fewer and fewer licenses during the 1990s over increasing concerns about India's missile and nuclear programs. India's decision to test a series of nuclear weapons in May 1998 ended any chances of a rapprochement during the Clinton administration.

Ironically, India's new nuclear might did not help it avoid a serious clash in the summer of 1999 in the Kargil portion of Kashmir. Moreover, its space program failed to warn the Indian military of Pakistani incursions into the area.[75] As Nair notes: "The Kargil conflict was witness to not a single spacecraft being brought into the conflict although India was an established space power by then," leading to a "lack of information in all aspects ranging from intelligence on enemy locations to targeting information, weather inputs, etc."[76] Existing satellites were unable to maintain continuous surveillance of the battlespace, thus revealing key gaps in India's military space program and accelerating the demand for new attention to its needs as well as an overall shift in national space priorities.[77]

9/11, GEORGE W. BUSH, AND THE U.S.-INDIAN RAPPROCHEMENT

The terrorist attacks on the United States in September 2001 ushered in a major strategic shift in U.S. policy toward India. Based on the common threat posed by Islamist-inspired terrorism and shared values of democracy, the Bush administration decided to back away from nonproliferation sanctions against India and open the path toward a new strategic relationship. After a visit by Prime Minister Vajpayee in November 2001, the two administrations released a list of priority areas for cooperation, which included space.[78] These plans resulted in the formation of the Indo-U.S. High Technology Cooperation Group in 2002, which in turn led to the Next Steps in Strategic Partnership (NSSP) agreement in January 2004.[79] The NSSP spelled out nuclear and space cooperation as two key areas for advancement, providing the groundwork for U.S. loosening of export controls for NASA cooperation in space science with ISRO. Although still limited in scope, the deal marked a major rapprochement for Indo-U.S. space relations and opened the way for substantive talks between ISRO and NASA, which led quickly to an agreement to fly U.S. experiments on the upcoming *Chandrayaan-1* lunar mission.

Benefiting from lower wages and an English-speaking workforce, investment by U.S. companies in the information technology (IT) sector accelerated as part of this rapprochement. Overall, U.S.-Indian trade surged 79 percent between 2000 and 2007,[80] with the Indian IT sector growing to 1.6 million employees, generating revenues of $60 billion a year and some 6 percent of India's gross domestic product.[81] This economic base provided stimulus to the space relationship as well, with American aerospace companies beginning to lobby the U.S. government for access to India's launch providers. The two sides also engaged in high-level military-to-military talks and, in the final days of the Bush administration, inked a controversial deal for cooperation in the nuclear energy sector, despite India's non-signature of the Non-Proliferation Treaty.

In the space field, the two space agencies working on the lunar mission faced significant hurdles from the U.S. export control system, as separate technology safeguards and technology-assistance agreements had to be put in place before any U.S. technology could be included in the mission, causing significant delays on the Indian side.[82] Indian officials balked at the use of the term "assistance" in the agreement—which smacked of colonialism—but eventually signed the required documents to allow the transfer of the U.S. payloads. A larger deal for Indian commercial cooperation with the Boeing corporation, however, was eventually abandoned because of U.S. bureaucratic obstacles. The success of this cooperative effort, particularly in flying a U.S. instrument that determined the presence of liquid water on the Moon's surface, provided excellent publicity for future joint missions.[83] Nevertheless, this new initiative remained limited to space science—not commercial space-launch or military sector activities.

India's economic opening after 2000 reached out to Asia as well. Particularly significant was the rapid acceleration in its trade with its rival China, with trade figures increasing from $3 billion to $18 billion in six years.[84] However, despite this veritable revolution in commerce, strategic relations remained cool. Space cooperation between the two sides remained nonexistent, in sharp contrast to India's growing space ties with Europe, the United States, and Japan.

CURRENT CIVIL AND COMMERCIAL SPACE ACTIVITIES

Space has become part of everyday life in India through the decades of work by ISRO. The public press covers key space developments closely, and acronyms like *INSAT-4CR* are known well enough by the average person to appear in newspaper headlines.[85] Fishermen carry receivers onboard to cue them with satellite-derived data on weather forecasts and the location of schools of fish. In these regards, India has succeeded in integrating space technology into the lives of a significant segment of its population and in making the public aware of these services.

As India has moved toward the commercialization of its space technology, it has achieved the greatest success to date in its space-launch industry, where it has benefited from its reputation for reliability and the relatively low cost of its services. In the past decade, India has developed a long list of clients, including Argentina, Belgium, Canada, Denmark, France, Germany, Indonesia, Japan, the Netherlands, Singapore, South Korea, Switzerland, and Turkey.[86] Moreover, it has sometimes offered flights to countries that have had difficulty finding other providers. For example, despite Iranian objections, India launched the Israeli military's *TecSar-1* reconnaissance satellite in 2008.[87] As part of the deal, India reportedly gained access to the use of *TecSar-1*'s high-resolution imagery.[88]

India's main launch vehicle today is the PSLV, whose medium-lift capabilities into polar orbits have been proven effective and are now marketed for low-Earth orbital payloads of 3,700 kilograms and eight hundred kilograms to geostationary transfer orbit.[89] Although it has recorded a number of successful flights with Russian upper stages, India is still perfecting its GSLV, which experienced problems in April 2010 when the main cryogenic engine on India's domestically produced third stage failed to ignite, thus preventing its satellite payload—the first element in India's planned GPS Aided Geo Augmented Navigational (GAGAN) network—from achieving orbit.[90]

India's Department of Space has focused in the past decade on commercializing space technologies developed by ISRO and marketing them both nationally and internationally through the Antrix Corporation. By 2007, this process had moved 270 technologies from the public to the private sector.[91] One area of success has been in direct-to-home television broadcasting, a market that has been growing at 20 to 30 percent a year and is expected to reach some twenty-five million households by 2013.[92]

The Indian space official Rajeev Lochan wrote in an essay published shortly after his death in 2008 that India's space infrastructure is critical to the country's economy, society, and environment.[93] For this reason, he argued, "The security of this infrastructure, its renewal and expansion as needed, and the uninterrupted and assured continuity of its operational services form the core of India's concerns about space security."[94] He also noted that space development cannot be part of a continued process of separating "have" from "have-not" portions of the national or global populations. Therefore, he posited that any meaningful definition of space security must include "Secure, sustainable, and denial-free access to and use of space *for peaceful purposes and for one and all* (italics in original)."[95]

One question that Lochan's points raise is whether ISRO's original mission to relieve poverty and promote national developmental aims is still being fulfilled. On this score, the record is mixed, according to Indian observers. In Mistry's evaluation, "in purely economic terms, India's space assets have cost more than the revenues they have generated."[96] In considering Nehru's original goals for the application of science to social needs, space technology does not seem to have lessened poverty in India substantially, as child malnutrition remains at 42.5 percent of the population (in comparison to 7 percent in China).[97] Thus, domestic challenges remain in deriving economic benefits from space that help the poorest Indians, even as the country moves to expand its space exploration and military space programs. But Mistry admits that India has gained a high degree of technological independence from other nations and benefits in "international prestige" and "foreign policy spin-offs."[98] While it is hard to quantify the value of these gains, support for the space program has been a political constant across a long series of Indian political administrations, thus suggesting their value to the country. In moving to include space exploration in its portfolio, according to the physicist Bharath Gopalalswamy, "India no longer views space as only enhancing the living conditions of its citizens but also as a measure of global prestige."[99] ISRO and NASA, for example, are continuing discussions to identify future missions of mutual interest, supported by significant high-level political backing in both capitals.

India's long-term space plans include considerable attention to Earth observation. The former ISRO chairman Krishnaswami Kasturirangan outlined India's strategy through 2025 as including the following priority spacecraft: *Resourcesat* (with an advanced multispectral camera), *Cartosats* (with improved high-resolution imaging and mapping capabilities), *Radar Imaging Satellite* (*Risat*) (for crop and moisture assessments), *Oceansat* (with a scatterometer for wind readings), and *WiFSSat* (a small satellite intended especially for disaster support and science-related missions).[100] India is also involved in a cooperative venture with Japan's JAXA and four other space agencies to build a microsatellite for Earth and ocean observation of the Asian region.[101]

India's commitment to pushing its space program forward has been proven to date by the government's strong budgetary support. As one analyst writes, "the total space budget allocated to India's Department of Space has been growing at a rate greater than any of the other major space faring countries," citing a 35 percent increase in the 2010 budget.[102] As the Indian analyst Ajey Lele observes, "Having Space technology is . . . becoming a symbol of development."[103] The drive for benefits from space commerce, he argues, "has led to a new form of Space Race, where every state is looking for a share."[104]

As in the United States and a number of other spacefaring countries, there is a challenge in making sure India has qualified personnel to staff its ambitious new missions. China clearly has the leverage to "push" graduates into the space program. India faces the threat that "most technically-minded Indians gravitate to the IT industry rather than to the space program."[105] One space official also admits, "The IT revolution has taken some of the brains from ISRO."[106] This will be a generic challenge to state-run space agencies in free societies in the coming decades. India's heavily bureaucratic educational system and slowness to allow foreign universities to offer programs in India has similarly hampered the development of skilled professionals for the space field, as well as other industries.[107]

INDIA'S NEW INTEREST IN MILITARY SPACE

India's military leaders, who see the past as having unnaturally excluded military activities, clearly welcome a role in space. In contrast to the U.S., Russian, or Chinese militaries, India's armed forces had virtually no role in day-to-day space operations before 2007.[108] Military training of specialized personnel in space operations was also minimal to nonexistent, since space systems in India were traditionally managed by civilians.[109]

But the 2007 Chinese ASAT test changed all of that by refocusing top-level Indian attention on military space capabilities after years of merely discussing the matter without taking any decision. In June 2008, India announced its formation of an Integrated Space Cell (under the Integrated Defence Services Headquarters) to coordinate India's military space activities with ISRO and begin the process of establishing operational command capabilities.[110] According to Dipankar Banerjee, India's formation of an aerospace command "is bound to impact on India's long-held policy of ensuring space is non-weaponized and used for peaceful purposes."[111]

Because of India's rising concerns with China's strategic modernization, Beijing's continuing assistance to Pakistan, and concerns about falling behind in precision guidance and deep-strike capabilities, India's military has put a new priority on the development of advanced space-based reconnaissance, navigational aids, targeting, early warning, communications, electronic intelligence, and possibly active defenses.

Beside the absence of a dedicated, military-purpose GPS system, India's armed forces have had to struggle to get adequate intelligence data from space-based platforms, such as those enjoyed by a number of modern militaries. Until the 2001 launch of the *IRS Technology Experiment Satellite*, India's military had lacked direct access to detailed photographic imagery from space.[112] Military experts called the coverage "grossly inadequate for Indian military requirements,"[113] causing them to resort to commercial purchases of imagery from U.S. remote-sensing companies and the Israeli government, among other sources. Today, the military continues to suffer from a "lack of institutional support"[114] for space activity, in comparison with ISRO and the various institutes under the civilian Department of Space. But India's *Cartosat-2A* satellite, with .8-meter optical resolution and Israeli synthetic-aperture radar technology, launched by ISRO in April 2008, has been transferred to military control and now serves the Indian Defence Intelligence Agency's Defence Imagery Processing and Analysis Centre.[115] In addition, the *Risat-2* satellite, with all-weather synthetic-aperture radar provided by Israel Aerospace Industries, achieved orbit in April 2009,[116] and a second optical-imaging satellite (*Cartosat-2B*) began transmitting data in July 2010.[117] The intelligence center also receives direct satellite feeds from Russian and Israeli reconnaissance assets.[118] Although planned primarily to serve agricultural and disaster management missions, four Indian Radar Imaging Satellites will also apparently be made available to military customers for specific all-weather intelligence requirements, as needed.[119]

To expand on these still-limited capabilities, the Indian military has issued a new doctrine called "Defence Space Vision 2020."[120] The new doctrine calls for a phased expansion, beginning with space communications and surveillance capabilities and the establishment of military-run operational capabilities under its new Integrated Space Cell.[121] Under the plan, the three branches of the Indian armed forces (navy, army, and air force) will each receive a dedicated satellite from ISRO by 2012.[122] Additional training activities, ground- and space-based hardware, and command/control capabilities will be developed in subsequent years.

Yet India faces some surprising gaps, given the longevity of its program. As the Indian legal scholar V. S. Mani writes, despite the decades-old existence of such key space bodies as the Department of Space and the Space Commission, "there is no broad-based legal framework to formalize their existence and operations."[123] Moreover, he observes: "the Government of India is yet to declare formally a coherently formulated space policy."[124]

India's Space Security Policies

Although India has ratified all of the major space treaties, questions have arisen recently about the depth of its commitment to them, despite decades of rhetoric regarding its peaceful intentions in space. As Mani observes: "no attempt has been made so far to enact any law to implement the obligations undertaken by India under the various treaties."[125]

Indian analysts identify China as their main space threat. But they seem to agree with Chinese space officials on the negative role of the United States in addressing space security issues, at least before 2009. As Lele argues: "Presently, [a] universally accepted space architecture appears to be a distant dream because of the stand taken by the US against any kind of treaty . . . that helps in stopping or limiting the weaponization of space."[126]

Yet the concept of an Asian space race comes clearly into play for India when it considers China's military space advantages and their implications on India's security. As Nair argues in addressing China's growing military space capabilities and their impact on Asia: "In such a volatile situation, a regional imbalance of power would only serve to exacerbate regional tensions, suspicions and rivalries and most likely trigger a space weaponization race in addition to the prevailing nuclear competitions and tensions."[127] This is the risk, he argues, of China's current course and, by implication, the failure of India to respond in order to "balance" the situation. However, such dynamics risk elevating Asia's current space race into the very arms race that all countries hope to avoid (or say they do).

In January 2010, Director General of India's Defence Research and Development Organisation (DRDO) V. K. Saraswat seemed to throw down a gauntlet to the Chinese, although without mentioning any country specifically. He openly admitted Indian efforts to develop an antisatellite capability, adding, "we are also working on how to deny the enemy access to its space assets."[128] India's military is clearly worried about China's prior development of a tested ASAT capability. As Air Chief Marshal P. V. Naik explained his concern over the recent activity of India's regional rival in space: "Our satellites are vulnerable to ASAT weapon systems because our neighborhood possesses one."[129] By 2011, DRDO's Saraswat had revised his statement about India's capabilities by claiming, "we have all the technology elements which are required to integrate a system . . . [to] defend our satellites" through the country's missile-defense program.[130]

At the same time, several Indian statements have asserted that India will not engage in kinetic testing of such a weapon, out of concerns about debris, raising questions about whether India's planned response might instead be a ground-based laser weapon. Several failures of India's kinetic missile-defense system from 2007 to 2010 also suggest that India has not mastered all of the technologies involved and may be testing a system in order to establish a precedent in case of future space arms control talks. As Victoria Samson observes in regard to India's past frustration with being left out as an original weapons state in the 1968 nuclear Non-Proliferation Treaty: "There are some within India who have taken that lesson to heart and want India to develop an ASAT capability so that India would be grandfathered in, should any future treaty or international agreement ban ASATs."[131] Of course, such a move would violate India's stated policy on the nonweaponization of space.

A recent surge in Indian defense spending to nearly $30 billion a year and a ramp-up in expenditures on big-ticket military items is likely to benefit space as well.[132] As it pursues the development of missile defenses, India has indicated that it plans to place an early-warning radar in geostationary orbit to detect launches and maximize the time available to begin its tracking and intercept sequence.[133] But it may be some time before India capitalizes on advanced, GPS-based targeting systems for the military. Nair argues that the GAGAN system that India plans to develop is still oriented toward civilian applications and currently has no planned military applications.[134]

In thinking about the future, Indian military analysts and commentators frequently point to the weaknesses of India's space segment and the risks this poses in regard to its neighbors. M. P. Anil Kumar, a retired Indian Air Force officer, worries about China's possible emergence as a "rogue space power" and its assistance to Pakistan.[135] He writes: "China is light years ahead of us in offensive space technology; so our endeavor should be 'space denial.' In [the] case of Pakistan, we must go all out to achieve total 'space control.'" Another observer, the former DRDO head V. K. Aatre, sees the extension of conflict into space as inevitable: "We have fought wars in the air, water and land. But the way things are going, Star wars will no longer be just a fiction."[136]

At the same time, it is important to note that such comments can be found among hard-line military analysts in China, Russia, and the United States as well. Nevertheless, the DRDO's "Technology Perspective and Capability Roadmap" issued in May 2010 specifically called for the development of ASAT weapons "for electronic or physical destruction of satellites," not only in low-Earth orbit but also, somewhat surprisingly, all the way up to the geostationary belt.[137] The key question will be whether India's civilian leadership will support such a move and whether India's geostrategic needs foster the development of offensive military space technology or merely military support functions from space.

SPACE, REGIONAL POLITICS, AND THE FUTURE

India continues to offer the world a different perspective on space, stemming from its former colonial status and its current status as a developing nation. Its traditional definition of space security has therefore differed from the military orientation taken early on by the Soviet Union and the United States. As the ISRO official Narayana Moorthi explains, the concept of "human security" applied to space "enables a broadening of our notion of space security from its traditional conception in military terms, to encompass other threats (including those emanating from poverty, lack of education, health hazards, environmental degradation and natural disasters)."[138] Similarly, Indian perspectives on space security also focus on the "distributional" dimension: that is, the notion that space is the "common heritage of mankind" and that benefits from space should be shared. However, India has not ratified a key treaty that calls for the distribution of benefits from development of lunar resources: the 1979 Moon Treaty. Thus, there remains a gap between India's rhetoric and its international policy.

During deliberations leading to the passage of the 2007 United Nations Debris Mitigation Guidelines, India initially appeared to be playing the role of a potential spoiler.[139] In light of past space practices by the Soviet Union and the United States, which had put the bulk of the then-extant orbital debris into orbit, Indian representatives argued that unfairly forcing India and other developing countries to abide by strict debris-mitigation guidelines now amounted to "cultural imperialism." Indian diplomats initially threatened to block agreement over the proposed guidelines.[140] In the end, after receiving promises of technology from the United States and other space powers, India altered its policy and allowed the guidelines to be approved by acclamation at the United Nations in December 2007.

India is also diverging from its traditional focus on Earth applications to get involved in the high-prestige race for space exploration firsts within Asia. India's *Chandrayaan-1* lunar mission brought new attention to India's space science efforts. The program marked the first time India had moved beyond Earth orbit to the celestial bodies. Equally significant, ISRO worked with a variety of foreign partners, even offering space on the mission for foreign experiments. This move both helped build India's ties with key international space programs and added scientific merit to the mission. In the end, *Chandrayaan-1* hosted payloads from German, British, and Swedish research institutes (through the European Space Agency), Bulgaria, and two U.S. payloads (one from NASA and another from a joint Brown University–Jet Propulsion Laboratory consortium).[141] This cooperation greatly increased the international attention paid to the flight and contributed to its general success.

But the mission also experienced difficulties, highlighting weaknesses in India's space technology and transparency. Because of an unexpected problem with the overheating of the *Chandrayaan-1* spacecraft, it had to be boosted into a higher orbit around the Moon, distorting its data readings. Yet ISRO refused to share this information with its foreign partners until July 2009 (nearly a year after the satellite's launch), leaving other space programs frustrated, particularly after the spacecraft experienced attitude-control problems and crashed into the Moon's surface only a month later. This experience suggests that India's national pride may, in some instances, get in the way of good science and effective cooperation with international partners. India also continues to seek technology transfer from its cooperative missions, which has become somewhat of an obstacle in its relations with NASA.

Although not to the same extent as China or Japan, India has also begun to use space to serve its broader foreign policy objectives with less-developed countries through the offer of cooperation. In Afghanistan, for example, India recently included remote satellite transmitters for acquiring space-based data in a $1.2 billion aid package to the Kabul government.[142] The Department of Space also recently created the Indian Institute of Space Science and Technology to offer both Indian and other students from the region undergraduate and graduate education in the space field.

Looking ahead, India has made a major commitment to human space-flight, for its high prestige value and the perceived negative implications of allowing China to dominate Asian spaceflight. It has budgeted some $2 billion to develop the first stages of an indigenous manned program and is seeking technical assistance from Russia for this effort through a deal reached in December 2008.[143] The plans include Russian training of an Indian cosmonaut for a Russian Soyuz flight in 2013 and then an Indian-only mission in 2015 in a GSLV-launched capsule capable of carrying three persons into low-Earth orbit.[144] One senior ISRO official has also mentioned a possible Moon landing by 2020.[145] As the space analyst Asif Siddiqi observes regarding India's motivations: "From a purely practical perspective, the manned program seems unnecessary to ISRO's original mandate; it is clear that the manned program is not about the pursuit of scientific or technical knowledge or about alleviating poverty—it is first and foremost about prestige."[146] Other high-profile missions include the *Chandrayaan-2* mission planned for 2013, with a lunar lander/rover (to be built with Russian assistance),[147] and a planned Mars-bound spacecraft in the 2016 to 2018 timeframe.[148]

Indian-U.S. space relations moved to a higher footing in November 2010, when the Obama administration announced during the president's trip to India that it was removing ISRO's main space centers from its list of organizations requiring a U.S. special license to engage in bilateral trade or other cooperation.[149] In addition, the two sides discussed a variety of space science missions, human spaceflight opportunities, and a possible future agreement to allow U.S. commercial space launches aboard Indian boosters.

CONCLUSION

India's progress in space has been marked by steady development and an emphasis on civilian applications. This unique focus, however, has begun to change as the country has emerged politically, economically, and militarily as a larger player on the regional and international scenes. India's expanding technical capabilities have also created opportunities for the space program to step up to the next level and begin to compete with other major spacefaring countries in previously neglected areas, such as space science and exploration. In addition, India seems to be motivated by China's advance into human spaceflight to develop its own man-rated boosters and manned programs, both for low-Earth orbit and the Moon. In its military programs, India has issued only slightly veiled warnings to China that it will seek to defend its assets and deny China the ability to use space forces against it in a future conflict. Whether India will be able to follow this rhetoric up and what damage it is willing to sustain to its previously peaceful image as a space actor remains to be seen.

With its growing economy, India has visions of a larger regional and international role for itself in the twenty-first century. It sees itself as having achieved recognition of its nuclear status, its critical role in the global economy, and its virtual alliance with the United States in promoting global democracy. For these reasons, as the former U.S. South Asian ambassador and official Teresita Schaffer argues: "New Delhi wants a greater voice in global governance."[150] But will such a role lead to greater cooperation or conflict with Beijing over space, much less Washington? India's voting patterns in the United Nations have coincided with those of the United States only 14 percent of the time, even less than the overall average worldwide of 18 percent.[151] As the Indian analyst Vadlamudi observes: "Political differences between India and the United States arising out of divergent geopolitical objectives in certain areas may influence the development and implementation of the next stages of the partnership."[152] As for China, as Evan Feigenbaum writes: "India does not view Beijing as an honest broker," and China "has begun to replace Pakistan at the center of Indian defense planning."[153] Yet, despite its differences with the world's two leading powers, it is not evident how or if India can offer an alternative as a regional or world leader, since it has thus far failed to offer specific "public goods"[154] such as security, economic benefits, or political leadership. Regarding space, it has also not yet offered new proposals for regional cooperation or for broader international space security.

Yet India has succeeded over the past two decades in establishing itself as a major new entity on the global scene in the areas of commercial space and, most recently, space science. Its forays into military space suggest a serious intention but not yet a robust capability. India seems to recognize that continuing to develop its military space technologies and simultaneously initiating a human spaceflight program is going to be expensive. But it faces an unpleasant alternative: relegating itself to falling further behind China in a field of technology that is critical to national defense and to its economic competitiveness. For these reasons, it seems that India is likely to pay the price, although unavoidable and rapidly approaching costs for modernizing India's vast and dilapidated domestic infrastructure and for dealing with enduring socioeconomic inequalities will sorely test its commitment.

Whether India will become a true leader in space is far from clear. The *Chandrayaan-1* mission marked a significant step in this direction, although the technologies it provided were not groundbreaking. India has also offered considerable rhetoric in the past about its desire to promote the nonweaponization of space. In the presence of stillborn international talks on this subject, India has the opportunity to step up and offer new ideas. It has not yet done so, and it may choose instead to follow the lead of the United States or other powers. The most serious missing element in its effort to expand its space presence has been any bridge building with China, despite the rapid growth of Sino-Indian commercial ties, which now boast $60 billion a year in bilateral trade.[155] Herein lies an irony that India will sooner or later have to face. It is motivated in significant measure by a desire to match and best China's space program and increasingly to defend its interests and assets in space. But an aggressive turn in its space policy is likely to harm its economy and stimulate Beijing to respond with its own military one-upmanship, thus putting India further behind and creating conditions where India's civilian space program may find it harder to conduct its activities. An editorial appearing in the trade weekly *Space News* captured this double-edged sword facing India's space program: "Should India decide to go through with a full-blown test of its anti-satellite capability," it "would inflame tensions and encourage others to conduct similar tests."[156] Thus, India faces a new security dilemma in space, even as it seeks to diversify its program and gain international recognition for its seat at the space table.

CHAPTER FIVE

THE SOUTH KOREAN SPACE PROGRAM

Emerging from Dependency

Although China, India, and Japan have received the bulk of recent attention for their space developments, the emergence of the Republic of Korea (ROK, or South Korea) as a new player has arguably been even more dynamic over the past two decades. From having virtually no space capability, organizational structure, or overall plan as late as the early 1990s, South Korea has engaged in a major effort to enter the arena of spacefaring nations. As Doo Hwan Kim argues, "Korea is attempting to achieve fast-track development in the space race," despite its late starting date.[1] Although its strategy focused initially on purchases of foreign technology, the ROK has increasingly sought to develop independent capabilities as a satellite producer, space services provider, and space-launching country. In this sense, South Korea is finally emerging from its long dependency on the United States for space-derived information. But Seoul's dual desire for both technological advancement and eventual autonomy has not prevented it from continuing to cooperate with the United States and with other countries to serve its purposes, including Russia, France, India, Israel, Japan, and even China.

In this regard, the ROK's position within the Asian space world has been perhaps the most politically "balanced" among the developed programs. It has sought to maintain ties with a range of countries both to push its technology forward and to prevent its possible isolation. Perhaps this behavior stems from an uncertainty about which way to turn, toward

China—its largest trading partner—or the United States, its main ally but a country that has denied it space and launcher technology for many years. These ties are now on the mend, but it is striking that Seoul only succeeded in reaching its first cooperative agreement with NASA in 2008, decades after similar U.S. arrangements with other allies (and even some adversaries, such as the Soviet Union/Russia). Still, in its current position, Seoul may represent a possible hub or "pivot"[2] for the future development of broader regional space cooperation.

South Korea is a smaller country in terms of size, finances, and population than other major Asian spacefaring nations. Lacking the resources of a large power, it has adopted a more integrative strategy concerning its Asian neighbors. As the space politics analyst Wade Huntley observes, the larger space programs "tend to see the other important powers as rivals, if not adversaries,"[3] whereas smaller countries like South Korea "look at the world differently."[4] For them, "the key is to develop relationships,"[5] since they are less capable of going it alone in overcoming the high entry costs to entering the space field.

Developing a space program has represented a major national challenge for South Korea, a country that was forced to rebuild itself and its economy both after World War II and the Korean War. Seoul has pursued the goal of space activity with considerable state-led investment and the rousing of nationalism, as it has in other fields seen as critical by the government. Although its attempts to orbit a satellite have thus far failed, its now-advanced satellite manufacturing capabilities and its experience in sending an astronaut to the *International Space Station* (*ISS*) in April 2008 have put it firmly on a course toward its objective of becoming one of the world's "top ten" space powers by 2015.[6] It is also creating an infrastructure to cement that place. Nevertheless, as the Korean analyst Kim Kyung-Min writes, "the notion of a 'space program' still sounds exotic and adventurous" to most South Koreans.[7] The country's rapid advance into the space field has left some wondering about the value of the program and others questioning whether these expenditures are sustainable amid current budgetary difficulties.

The motivations for Seoul's new focus on space can be found in a number of specific factors: (1) South Korea's economic development aims and its pattern of state-led industrial development in other critical sectors in the past; (2) the threats posed by North Korea's nuclear and missile developments and the ROK's need to have an independent monitoring capability for national security purposes; and (3) South Korean national pride and desire to be recognized as an independent, modern, and technologically advanced society. To support these aims, the ROK has benefited as a late developer from the availability of foreign partners and the training of a solid pool of Ph.D. scientists and engineers in the United States and other spacefaring countries, many of whom have returned home to pursue careers in ROK space science and commerce. The gleaming grounds and facilities of the Korea Aerospace Research Institute in Taejon, one of a number of high-tech government-supported centers there, highlight the seriousness of this investment and the country's strong commitment to succeed.

This chapter traces the history, politics, economic motivations, and defense-related aspects of South Korea's evolving space capabilities. It highlights the ROK's challenges as a late developer (such as the alliance-related constraints on its rocket development program), the role of the state in overcoming considerable hurdles, and its use of foreign technology to jumpstart its program. But South Korea has succeeded in making remarkable progress in a short period of time, thanks to its determination, governmental support, and exceptional organizational capabilities, validating certain predictions of both Veblen and Gerschenkron. Yet Seoul's ability to maintain a broad set of space capabilities, including a human spaceflight component, is already being challenged. Greater pressures on its future budget compared to that of some of its larger Asian neighbors and the competitiveness of its products, particularly after it enters into the global market as a provider of space technology and services, will be future tests of the ROK's skill and commitment.

ROOTS OF THE SOUTH KOREAN SPACE PROGRAM

The first challenge the South Korean government faced in regard to space was to establish the building blocks for the country's scientific-technological base. This task would take much of the country's first two decades of existence. During the Japanese colonial period, graduate education had taken place almost exclusively in Japan. Following Korea's liberation by Soviet troops in the summer of 1945 and the South's separation from the North with the arrival of U.S. occupation troops, the country struggled through post–World War II economic reconstruction amid considerable political disarray. Unfortunately, the South had inherited a territory never meant to be independent economically from the more industrialized north. As Young-Gul Kim summarizes, "The end of World War II left [South] Korea with very few industries in the modern sense and no technical personnel pool."[8] With the failure of political order to be established even after the Republic of Korea was formed in 1948 (under the long-time U.S.-based exile Syngman Rhee), little progress had been made in the economy by 1950, when the onset of the Korean War destroyed even this weak foundation.

The difficult post-1953 conditions only strengthened the U.S. role, relegating the ROK to a dependency position in economic, military, and broader security affairs. Immediately after the Korean War, South Korea's per capita gross domestic product (GDP) stood at a paltry $70 and its total national GDP at only $1.5 billion.[9] Widespread corruption, continued economic problems, and serious political opposition to the increasingly authoritarian Rhee government in the late 1950s led eventually to the establishment of military rule in May 1961 under Major General Park Chung-hee. Park's national development program focused on reconstruction and creation of an industrial base, in an environment characterized by external political threats from the North and repression of dissent at home. Space technology still represented a luxury the ROK had no time, personnel, or funds to pursue.

The Park government reorganized the state apparatus and began a series of five-year national economic plans in 1962 to jumpstart the economy with focused state investment and tough industrial discipline. As Young-Hwan Choi summarizes the next twenty years: "Having failed to ride the wave of the first Industrial Revolution . . . [South] Korea [underwent] a condensed industrial revolution in the latter half of the twentieth century."[10] Park, who had been trained under the Japanese military, adopted a state-led model based on large conglomerates, or *chaebol*, which would lead to early economic advantages but certain later problems.

With the United States providing security and training to the South Korean military and police forces, Washington exercised nearly complete control over the ROK's defense programs and investment. Indeed, during the 1960s, U.S. assistance constituted "50% of the national budget and more than 70% of defense expenditures."[11] In addition, before any forays could be made toward missile development, South Korea needed first to develop a cadre of qualified scientists and engineers. This meant creating and staffing specialized academic institutions. Regarding the status of ROK technical education at the time, Young-Gul Kim observes: "In 1969, there were only about 600 full-time graduate students in all science and engineering fields in all of South Korea," and these were scattered across many universities and departments.[12] Moreover, the quality of the programs was generally poor.

As the ROK's economy continued to recover under the planned economy initiated by President Park, educational reform began to emerge as a key governmental task. Although political unrest caused Park to declare martial law in 1972, the government went forward with the establishment of the Korea Advanced Institute of Science and Technology (KAIST) to expand postgraduate training in engineering and applied sciences.[13] The Park government exempted KAIST's students from the then-mandatory three-year military service term and paid all costs of the educational program.[14] In order to give KAIST latitude for the rapid development of a faculty and courses of study, new regulations removed it from the restrictive oversight of the Ministry of Education and placed it instead under the more forward-leaning Ministry of Science and Technology (MOST).

Meanwhile, concerns for South Korea's security, fear of a possible U.S. troop withdrawal under President Richard Nixon, and emerging North Korean missile activities caused President Park to initiate a surface-to-surface missile program in 1971 as well as to begin work toward a possible nuclear weapon.[15] U.S. compromises, threats, and carrots under both Presidents Nixon and Jimmy Carter eventually halted the nuclear effort, but the ROK continued its missile program. Technology, however, remained a serious hurdle. In 1979, in return for U.S. assistance in the conversion of U.S.-provided Nike-Hercules air-defense missiles to a surface-to-surface function, South Korea reluctantly agreed to limit the range of its future missiles to 180 kilometers—adequate to the task of attacking North Korean front-line forces but not far enough to reach Pyongyang, which Washington saw as potentially destabilizing.[16] This effort essentially tethered Seoul's missile ambitions to short-range defensive objectives only.[17] By 1982, the Ministry of National Defense had deployed one battery of the new missiles and decided to shut down the program's development team, given these constraints and other military options for halting a North Korean assault.[18]

Following the bloody North Korean terrorist attack on ROK officials in Burma in 1983, however, South Korea reconstituted its missile team and tasked it with the development of a maximum-range system (within the U.S. limits) with improved guidance technology.[19] Later, the team was applied to the task of building a small sounding rocket program. Eventually, it was hoped that this expertise might lead toward a space-launch capability. The number of skilled ROK technicians had now begun to grow enough to constitute a critical mass of expertise from which to expand. Indeed, by 1985, the annual number of M.S. and Ph.D. degrees awarded in South Korea in engineering and science had jumped to 7,502.[20] In fields where it was possible, the government's goal was to attempt to use technological purchases, reverse engineering, and improvements in education to accelerate the pace of national technical advancement. In manufacturing, the ROK government had generally pursued a strategy of purchasing technology from abroad rather than accepting direct investment and foreign ownership. As two experts observed in the late 1980s: "Korea has hired technology but has not hired production."[21] This route was more expensive, as South Korea had to pay higher costs for licensing, but it was achieving its objective.[22] Finally, space's turn had come. In 1985, MOST issued a ten-year scientific-technical plan that called for a series of specific space-related projects.

SOUTH KOREA'S INITIAL FORAYS INTO MISSILES AND SPACE (1980s TO LATE 1990s)

With the country beginning to move from military leadership to civilian control on the eve of the 1988 Olympics, the Chun Du-hwan government began to outline an economic rationale for moving the country into the field of space technology. In 1987, the ROK legislature approved the "Aerospace Industry Development and Promotion Act," which included the first significant funding for space projects. To accomplish these goals, the Roh Tae-woo government established the Korea Aerospace Research Institute (KARI) in 1989 under the Korean Institute of Machinery and Materials.[23]

By the early 1990s, South Korea had adopted an import-substitution strategy to work with European and American companies in order to gain know-how about satellite construction and related technologies for civilian purposes.[24] Its first satellite (built by KAIST in collaboration with Britain's

University of Surrey) reached orbit aboard a French launcher in 1992.[25] Called the *Korean Institute of Technology Satellite* (*Kitsat-1* or, in Korean, *Uribyol-1*), the spacecraft was soon followed by two more advanced small satellites in 1993 and 1999. These spacecraft carried increasingly sophisticated digital-imaging cameras and communication systems into low-Earth orbit, powered first by fixed and then (on *Kitsat-3*) deployable solar arrays.

In terms of launch vehicles, the early 1990s saw KARI emerge as the manager of the Korean Sounding Rocket (KSR) program, as the government had opted for civilian control to replace the former military effort. KARI launched the first KSR-1 in 1993, with a similar follow-up test later that year. This single-stage rocket used solid fuel to reach an altitude of seventy-five kilometers.[26] A two-stage KSR-2 rocket followed with tests in 1997 and 1998, reaching altitudes around 150 kilometers.[27] But, besides beginning to bump into its missile-range agreement with the United States, South Korea faced significant technical challenges in moving from solid-fuel military missiles to the more controllable liquid-fuel rockets optimal for civilian space-launch purposes.

Bolstered by this progress and with increasing pressure from interested scientists, South Korea issued its first devoted Basic Plan on Mid-to-Long-Term National Space Development in 1996, sketching out a series of goals for the coming five years.[28] This marked the beginning of a real national space program. The plan included priority areas such as technological development for communications and remote-sensing satellites, rocket research, and commercial applications. The guidelines focused on indigenous technology but did not rule out continued cooperation with outside partners, especially to improve Korean know-how and experience in operating more advanced systems that could not yet be built in the ROK.

South Korea began to advance toward new capabilities with the so-called Koreasat (or, in Korean, Mugunghwa) series of geostationary communications satellites in order to offer permanent satellite broadcasting services to a nationwide audience.[29] The ROK contracted with the U.S. Lockheed Martin corporation to build these satellites, with launches aboard Delta rockets in 1995 and 1996 and from a French Ariane booster in 1999.

Despite the rapid pace of improvements, problems in the domestic educational system continued to hamper South Korea's space development. Experts lamented that while ROK universities were churning out large numbers of scientists and engineers, the quality of their courses and the value of domestic degrees remained comparatively low.[30] Thus, many of the leading aerospace experts continued to seek out advanced degrees abroad, particularly in the United States. Rather than implement tougher accreditation standards, the government chose not to challenge strong cultural values tied to the expansion of domestic educational opportunities. As Young-Gul Kim comments on the factors leading to this stance: "Korea's culture emphasizes harmony in society. Critical evaluation of existing institutions is discouraged."[31]

In hopes of improving its competitiveness and preparing the country to become a viable economic actor on the global stage, the first civilian president, Kim Young Sam, undertook an ambitious globalization policy in 1993 and 1994.[32] Such policies aimed to build on his national priorities of democratization and political normalization, after the considerable turbulence of the mid-to-late 1980s. South Korea also engaged in a strategy of multilateralism, which included efforts to participate in international arms control agreements and nonproliferation efforts, and, in the United Nations, the country sought to move beyond the political confines of the U.S.-ROK alliance. In continuation of this policy, South Korea began by the late 1990s under President Kim Dae-jung to seek an end to its 1979 missile agreement with the United States and to win ROK entry into the Missile Technology Control Regime (MTCR).[33]

But South Korea's near economic collapse in 1998 and the government's difficult decision to accept International Monetary Fund bailout funds highlighted the failures of its economic policies. This spurred a more far-reaching reform movement that began to reorganize the economy away from the old *chaebol* structure and root out other factors (including corruption) that had led to inefficiencies. Such policies and the further opening of the domestic market were required as conditions of the loan arrangement, further stressing South Koreans and heightening perceived risks of foreign intervention into the economy.

THE EXPANSION OF ROK SPACE ACTIVITIES
(LATE 1990s TO THE PRESENT)

But national security concerns soon intervened to renew the ROK's emphasis on the development of space capabilities. In August 1998, North

Korea's Taepodong-1 missile test and attempted satellite launch shocked South Korea's system of national security, laying bare the country's fundamental reliance upon the United States for space-derived intelligence on its neighbor, its most serious national security threat. ROK military authorities had to wait to receive critical information from U.S. forces about the launch site's preparation as well as data about the flight itself. This was an increasingly uncomfortable position for the ROK, particularly given emerging political differences with the United States over aspects of the military alliance at this time, including rules governing the status-of-forces agreement and criminal cases involving U.S. soldiers. The test also exposed the South's risk of losing the race to become the first of the two Koreas to orbit a satellite.[34] Although the North Korean *Kwangmyongsong-I* satellite failed to achieve orbit,[35] this surprise challenge to South Korea's technological position represented a wake-up call that Seoul was no longer alone in its space ambitions. Furthermore, even the less-than-successful Taepodong test (given its failed third stage) marked a major range improvement (1,300 kilometers) over the previous North Korean Nodong missile (with a tested range of only five hundred kilometers),[36] indicating a much-expanded military capability for the North, while the South remained constrained by the 180-kilometer limit imposed by the United States. Unlike the North's missiles, South Korea's short-range systems had no prospect of placing a satellite into orbit. All of these factors pushed satellite and rocket-related research to a higher priority under President Kim Dae-Jung (1998–2003) and President Roh Moo-Hyun (2003–2008).

Within the military, the Korean Air Force created a Space Weaponry Branch in the Air Force Studies and Analyses Wing to begin planning for possible space operations.[37] To address ROK weaknesses in Earth remote-sensing capabilities, KARI began work on the so-called *Korean Multi-Purpose Satellite* (*Kompsat* or, in Korean, *Arirang*) in cooperation with the U.S. company TRW. *Kompsat-1* entered orbit after a December 1999 launch from Vandenberg Air Force Base. The satellite had an electro-optical camera capable of 6.6-meter resolution and a series of other particle-sampling and ocean-surveillance sensors.[38] It represented South Korea's first step toward an independent Earth-imaging capability suitable for civilian remote sensing and future military reconnaissance. KARI next developed the more advanced *Kompsat-2* in cooperation with Israel's Elbit Systems Electro-Optical, Ltd., which provided more detailed images: one-meter resolution in black and white and four-meters in color.[39] The ROK contracted with Russia to place the satellite into low-Earth orbit in 2006.

KAIST had also initiated a science-oriented series of microsatellites (weighing approximately one hundred kilograms). This successor to the Kitsat program began in 1998 and aimed at development of space physics sensors and light monitoring technologies for Earth and atmospheric monitoring.[40] The first *Science and Technology Satellite (STSat)-1* entered low-Earth orbit in 2003 aboard a Russian booster,[41] with a follow-up spacecraft (*STSat-2*) already on the drawing board.

After further research and development activity, KARI scientists succeeded in 2000 in launching the considerably larger and more sophisticated KSR-3, a liquid-fuel rocket with a cryogenic oxidizer to an altitude of forty-three kilometers and a distance of eighty kilometers,[42] setting the stage for its planned Korea Space Launch Vehicle (KSLV). But, given the expected need for additional foreign technology, South Korea had already begun investigating potential entry into the MTCR. These negotiations finally succeeded in March 2001 in gaining ROK membership into this supplier organization in return for South Korean acceptance of a ban on deployment of military missiles with ranges greater than three hundred kilometers and a payload capacity exceeding five hundred kilograms. Benefitting from a 1995 MTCR policy shift made to attract Brazil into the agreement, the ROK received a similar waiver to allow Seoul to purchase space-launch technology from other MTCR members.[43] With this obstacle overcome, the ROK government announced in 2001 a $4.26 billion investment in its recently passed, multiyear "National Space Program."[44]

As one of the plan's top priorities, the KSLV project aimed to develop the ROK's first domestic launch vehicle capable of putting a microsatellite into low-Earth orbit. Starting in 2002, the South Korean government turned to the United States for possible assistance. But, contrary to the ROK's expectations, the United States was unreceptive. Tightened U.S. International Traffic in Arms Regulations (ITAR) regulations after 1999 and other restrictions in the U.S. Space Transportation Policy (meant to discourage the creation of competitors to U.S. launchers) made U.S. companies unwilling or unable to complete a deal with South Korea for delivery of a rocket stage. Instead, KARI turned to Russia's Khrunichev Design Bureau, a company hit hard by the Russian recession of the 1990s. It was only too eager to assist in the now-MTCR-approved ROK space program.

After a series of negotiations with Moscow beginning in 1998, the ROK and Russia agreed to terms in September 2004.[45] The $200 million pact included terms for the transfer of the Russian Angara-derived booster and cooperation in the construction of the launch facility at the Naro Space Center.[46] Reportedly, the first stone used in the construction at the Naro complex was brought by the Russians from the Baikonur Cosmodrome in Kazakhstan, from which Yuri Gagarin and many other cosmonauts had been launched into space. South Korea also paid some $22 million to Russia for the training of two astronauts, one of whom would fly to the *ISS* as part of a future Russian mission. South Korea ratified the Russian space deal in December 2006, and it officially entered into force in July 2007 after the Russian Duma's approval.[47] But Seoul now had to accept MTCR restrictions to use the Russian technology only for nonmilitary purposes and not to transfer information regarding the Russian systems to third countries.[48]

The Russian agreement also assisted Seoul in its vigorous pursuit of an astronaut program. The government sponsored a nationwide competition to identify appropriate candidates, eventually whittling down a list of 36,206 applicants.[49] The spacecraft would be piloted by Russian cosmonauts, so flight skills were not a requirement. The high-publicity competition eventually yielded two finalists—a thirty-year-old male computer scientist and former sportsman (San Ko) and a twenty-nine-year-old female graduate student in biotechnology (So-yeon Yi).[50]

Both candidates soon found themselves in Russia's Star City training for the upcoming mission to the *ISS*, with Ko as the designated front runner for the flight and Yi as the alternate. However, events turned out differently. Ko violated Russian regulations by removing training manuals from the cosmonaut center, taking one to South Korea before he returned it and reviewing another manual for pilots that he was not authorized to see.[51] Once the Russians found out, they removed him from the mission, and Yi moved into his slot. The incident proved embarrassing to the South Korean side.

Interviews conducted with Yi before her spaceflight, however, indicated the important symbolic value of the flight to South Korea: Yi's own hopes were that the peninsula would eventually be reunited and that the mission would be seen as an accomplishment for all Koreans.[52] The flight was a success and was covered extensively by the South Korean media, making

Yi a national hero. The mission won particular support from young Korean women, who viewed the flight as reaffirming their own potential in still-patriarchal South Korean society. Subsequent to Yi's highly publicized mission, overall public support for the human spaceflight program surged to an 87 percent approval rating.[53]

South Korea's subsequent efforts to launch the KSLV-1 and a satellite during the summer of 2009, however, met with failure. The original plan called for the Russian shipment of the first stage of the KSLV to be mated to an ROK-built second stage for a scheduled launch in 2007. This date eventually slipped to August 2009, because of technical delays on both sides. In the end, the rocket's two stages on the renamed Naro-1 both operated successfully, but the faring surrounding the combined KARI/KAIST/Gwangju Institute of Science and Technology *STSat-2* failed to release properly, causing the satellite to lose velocity and altitude and eventually fall back into the atmosphere.[54] The pieces were later recovered near Darwin, Australia. Although such initial failures are not uncommon in space programs, the experience marked an embarrassment for KARI and raised the chances of a successful North Korean satellite flight before the next launch could be attempted. Indeed, Pyongyang attempted to place a satellite into orbit in April 2010 but failed. A second South Korean attempt in June 2010 resulted in the explosion of the KSLV-1 about two minutes into the flight, resulting in some quiet finger pointing between the Russians and South Koreans as to who was at fault and an internal investigation by KARI.[55]

CURRENT DOMESTIC FACTORS IN SOUTH KOREA'S SPACE PROGRAM

South Korea has followed a clear space strategy of using technological advancements already made internationally to benefit its program and speed its path toward becoming a top-ten space power. Yet it continues to face challenges. Despite considerable progress, South Korea remains behind its major Asian space rivals in the critical areas of training and education. As one study observed in comparing Seoul's position to China's in thirteen key technology fields: "Korea is relatively strong in IT-related areas but weaker in aerospace."[56] The RAND Corporation estimates that South Korea lags behind China by about four years overall in terms of space technology,[57] putting part of the blame on the fact that "graduate-level education in Korea is [still] quite underdeveloped," with particular weaknesses in basic science and innovation.[58] The ROK fared better in commercial applications.

The South Korean government is aware of these challenges and is following them closely, with the aim of overcoming them. As KARI President Lee Joo-jin stated in October 2009: "In the medium and low altitude satellite fields, the country has reached 80 percent capability vis-à-vis leading countries like the United States and Russia, while the gap stands at roughly 40 percent for geosynchronous and communication satellites."[59] But this situation poses long-term questions about the sustainability of South Korea's recent surge into the space field and its future competitiveness in key areas such as innovation. Accordingly, KARI has recently opened a new Research Center for Satellite Information and an expanded Satellite Test and Integration Center, both aimed at increasingly domestic technical capabilities. As a KARI statement emphasizes, "this infrastructure has been developed using domestic technology, which means Korea has now attained international competitiveness to export Korean-made satellites as well as test facilities, [and] large test equipment under plant contracts to countries undertaking satellite development."[60] To date, only one South Korean private manufacturer (Satrec Initiative)—formed by a group of former KAIST engineers—has completed commercial contracts.[61] The company built a satellite bus for Malaysia's *RazakSAT*, which was orbited successfully aboard a U.S. Falcon-1 rocket in 2009, and another for the United Arab Emirates' *DubaiSat-1*, which was launched successfully by Russia in May 2009.[62] Satrec Initiative hopes to target Singapore and Turkey as possible future clients and is already working with UAE engineers to design and build a more advanced follow-up, *DubaiSat-2*, due in 2012.[63]

A potential limiting factor for South Korea's civil space program is its budget. From an initial staff of only thirty-five and a $3 million budget in 1989, KARI has grown to 667 personnel and a budget of $256 million.[64] Additional governmental support is provided to other space research institutions, such as KAIST (with its Satellite Technology Research Center), and to the ROK military. But funding has been flat in recent years, and the overall size of South Korea's space cadre is dwarfed by the number of space scientists and engineers in China, India, and Japan. The impact of the 2008 global crisis also hit the ROK harder than larger countries like China and

Russia, because the country lacked either natural resources or a large do-mestic market to fall back on. If space becomes identified as a luxury item in future years, the program's budget could face additional pressures.

One problem is in the area of human spaceflight. In the wake of Yi So-yeon's flight, KARI issued an optimistic assessment of this program and its goals:

> In addition to the advancement of Korea's international status through se-curing manned space technology as well as the creation of economic value from the far-reaching ripple effects throughout industrial circles, this [human spaceflight] program is an industry with very important social and cultural significance. This program inspires national self-respect and pride into all Koreans and provides an opportunity for the younger generation to foster a dream of scientific technology.[65]

However, budgetary realities have forced KARI to put the program on hold. Currently, no new astronauts are being trained for future flights.

One priority for technical development was an independent weather satellite. The geostationary *Communication, Ocean, and Meteorological Satellite (COMS)-I*, (or, in Korean, *Cheollian*), which was contracted in 2005 for joint production with the European EADS Astrium corporation, successfully reached orbit in June 2010 aboard a French Ariane rocket. It now provides South Korea with its first devoted oceanographic and meteo-rological data and enhances its communications network.[66] Korean news services noted that the satellite was only the seventh "national" weather satellite in the world.[67] A follow-up *COMS-2* is planned for 2014, to begin development of a nationwide satellite navigation augmentation system.[68] The program will also benefit from South Korean participation in Europe's Galileo precision satellite navigation network. Of course, all of these pro-grams will be expensive, since they involve foreign technology.

In terms of process, the KARI budget works on a yearly plan where sci-entific teams compete for funds allotted by the government on a project-by-project basis.[69] It is a time-consuming process that involves complex applications and then a process of political consent. Indeed, as KARI tries to expand, it faces a problem related to the absence of reliable institutional support for its staff and infrastructure. Despite the organization's rapid growth in recent years, less than 10 percent of KARI's yearly budget comes in the form of unrestricted institutional support from the government. Approximately 90 percent is tied to specific yearly research requests, and thus KARI is in a constant struggle for funding.

To help spur interest in space development, KARI and the city of Taejon hosted the three-thousand-person International Astronautical Congress in October 2009. The meeting provided a surge of interest and media coverage of South Korean space accomplishments and helped facilitate new ties with foreign aerospace companies and organizations. The conference's president, Berndt Feuerbacher of Germany, observed that Seoul's potential to advance in future space activity is likely to be a factor of how well it manages the linkage between "lightweight IT and space technology."[70] He concluded, "If this happens, the country can succeed."[71] To date, its progress has been rapid, but questions about its international competitiveness and market niche remain to be answered.

SECURITY CONSIDERATIONS

As North Korea's missile and nuclear programs advance, South Korea faces internal and geostrategic pressures to expand its military space effort. This security-related agenda may put it into conflict with the United States, China, Japan, and other space powers, depending on the ROK's direction and specific programs. In mid-2009, senior U.S. military officials agreed to begin talks with South Korea on a possible revision of MTCR restrictions to allow longer-range missiles,[72] but such talks have now been delayed repeatedly. As in its complex negotiations over proposed South Korean reprocessing of nuclear fuel, Washington does not want to set a precedent that will make it harder to "walk back" North Korean capabilities in the context of the Six-Party Talks. However, reports suggest that South Korea may already have a 1,500 kilometer-range cruise missile capable of hitting any site in the North.[73]

In regard to weapons for possible use against space, the South Korean Navy has had operational Aegis missile-defense systems aboard the destroyers *Sejong the Great* and *Yulgok Yi* since November 2008.[74] While not a devoted antispace or antisatellite system, the Aegis system's use by the U.S. Navy to destroy the *U.S. 193* satellite in February 2008 suggests that South Korea could conceivably conduct a similar upgrade. However, it would likely be very difficult, and South Korea has announced no such plans.[75] The ROK military is conducting work on truck-mounted, mobile lasers for use in antimissile and antiartillery defense.[76] An upgraded laser weapon could possibly be used for antisatellite purposes, although there is no evidence yet of such a system or related intentions by South Korea.

In the meantime, South Korea has begun to build an infrastructure for military support operations using space. In 2006, its *Koreasat-5* entered orbit to provide the first devoted military communications from geostationary orbit for the ROK armed forces.[77] South Korea contracted with France's Alcatel (now Thales Alenia) to build *Koreasat-5*, which also serves some civilian broadcasting functions.[78] A more advanced *Koreasat-6*, with a fifteen-year lifespan and built by France's Thales Alenia Space group and the U.S.-based Orbital Sciences Corporation, reached orbit in December 2010 to provide Ku-band communications for direct broadcasting and fixed satellite services.[79] Unlike in Japan, there has never been any formal restriction on military uses of space, and the ROK Agency for Defense Development has operated a dedicated facility in Taejon since 2005 to control the country's military space communications.[80] In addition, in 2007, the Korean Air Force established a Space Development Branch and began sending midcareer officers to Colorado Springs, for training at the U.S. Air Force's National Security Space Institute.[81]

To strengthen ROK reconnaissance capabilities, KARI plans to launch the dual-use *Kompsat-5* aboard a Russian rocket in late 2011, carrying Korea's first synthetic-aperture radar with a one-meter panchromatic resolution.[82] Also in line for the coming few years is the high-resolution, optical-imaging *Kompsat-3* (with .7-meter panchromatic resolution and 3.2 meters in color, built in cooperation with France's EADS Astrium), now scheduled for launch aboard a Japanese H-IIA.[83] Consistent with these programs, the ROK may well decide to emphasize passive military uses of space and stop short of using space for active defenses.

According to the Korean Peninsula analyst Daniel Pinkston, "South Korea is very sensitive about the implications of military encroachment in space. In the case of an arms race in space, Seoul could not compete in such a race and would feel extremely uncomfortable in the position of having to choose sides in a superpower [U.S.-Russian-Chinese] space conflict."[84] Unlike the United States, China, India, and North Korea, South Korea has pledged itself to remain a non-nuclear-weapon state through its membership in the Non-Proliferation Treaty and its signing of the International Atomic Energy Agency's Additional Protocol, allowing the inspectors access to all South Korean facilities. Therefore, South Korea is attempting to establish a different identity and reputation for itself—a leader in *peaceful* nuclear applications (particularly, energy) and a leading commercial player in other advanced technology fields, such as space. Yet, like Japan, it is also trying to maintain a strong and positive relationship with the United States, even as it asserts greater independence, and it has unique security needs given its northern neighbor. Thus, it seeks both to collaborate with NASA and other U.S. entities in space but also to emerge as an actor capable of developing its own scientific projects, offering its own space services on the international market, and possessing independent military support capabilities.

In terms of South Korea's strategic plans for itself within Asia, Seoul seems to be aiming at a "balanced" position between possible Chinese- and U.S.-led security factions. As Gilbert Rozman argues, "Similar to Italy, South Korea champions regionalism and balance within a region conducive to the voices of many rising above the voice of one or two."[85] Rozman notes South Korean efforts to woo Japan from too close an alignment with the United States to prevent its own potential abandonment in the middle in case of a possible U.S./Japan versus China split. But many Koreans still harbor strong anti-Japanese feelings stemming from imperial Japan's harsh occupation of the peninsula from 1910 to 1945.[86] Moreover, at least until reunification occurs with the North, all senior South Korean officials agree that the security alliance with the United States remains critical. Still, ROK-Japanese space relations are growing nonetheless.

Attitudes toward the United States have shifted back and forth since the emergence of democracy in 1988. Following the strongly pro-U.S. attitude of the ROK governments during the years of military rule, a generally pro–North Korean attitude emerged during the Kim Dae-Jung and the Roh Moo-Hyun governments.[87] Since 2006, however, the pendulum has shifted back toward a more pro-U.S. stance, as North Korea's testing of two nuclear weapons and multiple missiles has caused many South Koreans to change their minds about the North's intentions and, accordingly, the value of the U.S.-ROK alliance.

In its policy on space security, Seoul has navigated a careful line between the United States and China. ROK Ambassador to the Conference on Disarmament (CD) Chang Dong-hee exemplified this line in a speech in August 2008:

> The ROK, as one of the countries which is actively pursuing a peaceful space program, views space security, including the prevention of an arms race in outer space (PAROS) as an important issue of great relevance for the CD. In this regard, we welcome the draft PPWT [Prevention of the Placement of Weapons in Space Treaty], which was submitted by the Russian Federation and China this year, as a meaningful input to the substantive discussions in the CD on the agenda item of PAROS. We also believe [U.S.-backed] transparency and confidence-building measures (TCBMs) are highly important elements in building multilateral cooperation for the peaceful uses of outer space, and those two approaches of PPWT and TCBM can complement each other in obtaining the goal of peaceful uses of outer space.[88]

A question looking forward is whether Seoul will be able to maintain this middle ground or if it will be forced to lean toward Beijing/Moscow or Washington/Tokyo.

INTERNATIONAL PRESTIGE AND COOPERATION

South Korea continues to be motivated by the joint but sometimes contradictory goals of building its prestige and promoting international cooperation. Developing the ability to field indigenous space capabilities also remains an eventual important goal for many Koreans for boosting the national image abroad and building "soft power." Indeed, critics of So-yeon Yi's flight to the *ISS* pointed out that some of the luster of this spaceflight "first" for the Korean nation had been lost because it had taken place aboard a Russian spacecraft.[89] Thus, certain types of prestige-oriented programs pursued by KARI can represent a double-edged sword.

Yet using foreign programs and launchers has helped KARI make progress toward its own launchers, ROK satellite capabilities, and ROK space science and exploration programs. KARI's current policies are clearly geared toward expanding South Korea's international space network and the acquisition of building-block technologies for the future. Besides its cooperation with the United States, its participation in the *ISS* project and lunar network, its satellite purchases from France, and cooperation with the European Union's Galileo GPS project, as the Korean space analyst Kyung-Min Kim notes: "Bilateral cooperation with such nations as Russia, India, and Ukraine will also step up South Korea's efforts to acquire space technology."[90]

KARI's main international partnerships have depended on the type of technology involved.[91] In the area of civilian observation satellites, France's EADS is now the main partner, having replaced the U.S. TRW company out of cost concerns and technology-transfer issues. In the space-launch field, export control regulations and high costs have prevented deals with both U.S. and European companies. By contrast, Russia proved to be a willing partner. For the South Korean side, the intention regarding Russia from the start was a one-time arrangement to establish the ROK space-launch program and to facilitate the development of a larger, 100 percent Korean launcher, the KSLV-2. But current circumstances suggest that additional purchases from Russia to facilitate a successful flight for the Naro-1 will be necessary. Other partners have provided key satellite technologies and know-how. On a scientific level, KAIST benefited early on from cooperation with the United Kingdom on *Kitsat-1* and *-2* and, more recently, from work with a South African university in regard to the multispectral camera used on *Kitsat-3*.[92]

KARI has signed seven agreements with Japan since 1989 and has held bilateral space agency meetings yearly since 2005.[93] It also participates in the conferences and data exchanges of the Japanese-led Asia-Pacific Regional Space Agency Forum (APRSAF). With China, KARI has cooperated in small projects and meetings with the Chinese-led Asia Pacific Multilateral Cooperation in Space Technology and Applications (or AP-MCSTA) process since 1994, while also engaging less-developed space powers such as Mongolia and Thailand.[94] South Korea is part of the U.S.-led Global Exploration Strategy for a return to the Moon and is a member of the International Lunar Network, which is establishing standards for cooperation and data exchange among participants in lunar exploration. However, despite these prospective efforts, current ROK operational cooperation with NASA remains surprisingly limited when compared to U.S. joint activities in space with the European Space Agency (ESA), Japan's JAXA, and even Russia's Roskosmos. While prospective efforts may bear fruit in regard to lunar projects, such cooperation has mainly involved exchanges about small satellites at the NASA Ames Research Center and periodic meetings between officials from

NASA headquarters and KARI. These ties have yet to develop significant operational cooperation for actual space activities.[95] Highly restrictive U.S. export control regulations that govern not just missile technology but all space technology also constitute a continuing obstacle. These constraints have pushed South Korea in recent years toward ESA and Russia.

For this reason, Pinkston argues that "Washington risks being marginalized from South Korean projects in the future."[96] The result, according to Pinkston, is "the possibility that South Korea would develop more cooperative space relations with other space powers such as Russia and China, or even Pyongyang, which would increase the likelihood of Seoul siding with Beijing or Moscow in any future space disputes with Washington."[97] While the 2008 NASA-KARI agreement has begun to mend this potential rift, Seoul does have options in space and is not averse to exercising them to pursue its own interests. As Huntley observes regarding the possible purpose of pursuing space cooperation at several different levels, "The development of structured relationships provides assurances to smaller states that their interests are acknowledged and will continue to be served over time."[98] South Korea seems to fit this pattern.

Looking ahead, KARI has a series of projects in the pipeline, including a *STSat-3* space science satellite and the *Kompsat-3A* Earth-observation satellite (offering a suite of infrared sensors in a polar orbit at an altitude of 685 kilometers),[99] with the aim of expanding both its exploration program and its regional reputation in space science. In the long term, KARI expects to launch its own lunar orbiter by 2021 and a robotic lunar lander by 2026, thus joining a growing group of advanced space powers with active lunar exploration and research programs.[100]

In seeking independent access to space, KARI plans to complete its domestically produced KSLV-2 (or Naro-2) rocket by 2019[101] and to begin offering commercial launches from its Naro Space Center around 2020.[102] How such services fit within the international space market and whether there will be adequate demand remain to be seen.

CONCLUSION

South Korea has made great strides in space activity over the past twenty years. It has arguably already established itself as one of the top ten space countries by virtue of its satellite program and growing experience in managing and operating complex orbital systems, with capabilities nearing world standards. Yet South Korea is still swimming in deep waters if it expects to compete with its larger and wealthier rivals in Asia. This seems to have motivated South Korea to take a more flexible approach to partnering with other countries, as has been seen in its foreign space ties to date. Key questions remain, however, as to whether KARI will be successful in sustaining its remarkable recent progress. The satellite market remains highly competitive, as does the international launch-services market. Japan's difficulty in moving into both of these fields provides a sobering lesson for South Korea. But Seoul seems to be focusing on a different market, particularly those cash-strapped countries that may be attracted to its lower-priced space products. Yet the willingness and ability of the ROK government to continue its investment in the space program and its marketing abroad will be an important test of its commitment to find a place among its larger rivals and become established as a reliable provider.

Space technology also plays an increasing practical role for South Korea's domestic economy—which makes extensive use of satellite communications and navigation and draws increasingly on Earth remote-sensing data for a variety of commercial and military purposes. In Huntley's terms, Seoul also still wants a "seat at the table" for symbolic purposes—to prove to the North, its own citizens, and the international community that it has "arrived" as an advanced, technological nation. Just as with South Korea's initially difficult entry into the U.S. automobile market (with award-winning, high-quality vehicles eventually replacing the cheap and relatively unreliable early models), the ROK's space program has been taken on as a national challenge. Overcoming U.S. and other foreign export controls and other obstacles has become a major focus and even "cause" of the ROK government in its drive for both success and, at its roots, respect.

The future of South Korea's space program will hinge both on government funding decisions and on its relations with foreign partners. It is unlikely that Seoul will be able to become fully autonomous in space technology across the board. Indeed, such a strategy is overly optimistic and might need to be altered over time if the ROK is going to remain competitive as a niche player and thereby able to cooperate with leading programs as a peer. Its economy and resource base will make it difficult for Seoul to catch up and maintain a competitive position in all space-related areas; the

ROK is the proverbial fish swimming with whales. One wild card that remains in the deck, however, is the relationship to North Korea. If reunification comes within the next decade, the ROK space program could gain a significant step up, by gaining access to North Korean rocket technology. It could also begin to economize in various areas of manufacturing, by acquiring access to rare minerals and other useful raw materials. Still, even a unified Korea will face tremendous challenges, and at least a decade will be required to cover the costs of linking the two economies.

In the meantime, Seoul will be forced to "run faster" than its Asian competitors in space. For this reason, those elements of South Korea's current space strategy aimed at integration, cooperation, and efforts to prevent the emergence of aggressive foreign military activities seem most likely to serve its interests as a newly capable "middle" space power within Asia.

CHAPTER SIX

EMERGING ASIAN SPACE PROGRAMS

Australia, Indonesia, Malaysia, North Korea, Pakistan, the Philippines, Singapore, Taiwan, Thailand, and Vietnam

There is no set definition of what constitutes a "space program." Relevant measures include a range of possible criteria: using space-derived data, designing an experiment to fly on someone else's spacecraft, purchasing a satellite, building one's own spacecraft, operating a spacecraft in orbit (whether one's own or not), launching a spacecraft, having an astronaut go into space, or sending and returning astronauts into and from space with one's own technology. Given the lack of a definitive threshold, however, the issue is better viewed as a continuum starting with possession of some space-related capability and ranging on the other end to the possession of a full spectrum of civil, commercial, and military space assets. In selecting Japan, China, India, and South Korea as Asia's leading space actors, this book has focused on the technological sophistication and breadth of each country's space infrastructure and its production capabilities across these criteria, which collectively set these countries apart from the rest of Asia, at least currently. They are also the actors most cognizant of the Asian space race and most concerned about their places relative to the other major players. But other Asian countries are becoming increasingly active in space too. All are paying attention to the activities of their own regional rivals. Most often, however, these currently second-tier space countries are pursuing hedging strategies of a sort, and they are interested in benefiting from space information while hoping to slow the race forward by more developed actors (particularly in the military sector), thus ensuring that the first-tier countries do not either dominate space or ruin it for others.

A key advantage of these later-developing Asian space players is that they have an even greater ability to draw on the widespread range of space technologies now available on the international market, from GPS coordinates to weather data to space debris information to communications networks to remote-sensing images. All operate ground stations to receive foreign satellite data, some have operated foreign satellites, several have built and operated their own spacecraft, and a few have constructed rockets and attempted space launches of their own. A few have even sponsored astronauts that have received training elsewhere and conducted missions in space. Although their expenditures and breadth of capabilities are more limited than those of Asia's leading space powers, space technology has already begun to change their economies, societies, and militaries. For this reason, they are grouped together here as "emerging" Asian space powers.

This chapter focuses on ten countries with such developing space programs, each with a different focus, set of capabilities, relationship with foreign partners, and national space ambitions: Australia, Indonesia, Malaysia, North Korea, Pakistan, the Philippines, Singapore, Taiwan, Thailand, and Vietnam. A number of these countries have the potential to become more significant space players. Indeed, this is a dynamic situation, and a snapshot taken a decade from now would undoubtedly witness an increase in their absolute capabilities and perhaps their relative standing in regard to others in Asia, depending on the speed of each country's development and the decisions made by their national leaderships about how actively to pursue space-related versus other technology. None of these countries is standing still.

While each national space program is unique—and some have grander ambitions than others—a number of them can be said to share a small- or middle-country perspective on space. In other words, many of these capitals recognize that their resource bases will never allow them to match the larger space powers, and what they seek from space is the selective use of space assets to provide benefits to their societies, economies, and militaries. This involves using more cheaply acquired foreign technologies and services in most areas and developing domestic means in limited areas of particular national interest.

But size and scale are not necessarily the best measures for comparing these Asian space actors with the more advanced space powers. In many cases, the aims of these emerging actors are simply different. As Wade Huntley cogently observes: "Grasping how smaller powers are not just weaker great powers is essential to appreciating how their interests are formed and pursued—and accordingly, the nature of the function of power and governance in space."[1] In considering the challenges of regional space cooperation in Asia, understanding the individual interests of these countries is therefore important as well. Many of them are more open to cooperation than are some of the larger space actors discussed earlier in this book. Typically, they are also eager to prevent international deployment of military space capabilities, particularly space weapons, that they themselves are unlikely to be able to develop. In many cases, these peaceful policies are more realistic than idealistic in motivation and are rooted largely in financial and strategic limitations. Yet even those with small militaries numerically (such as Australia and Singapore) may incorporate advanced space technologies to promote their national security, though usually via passive means (such as data collection). Moreover, a few countries in this group are highly competitive or face such serious security concerns regarding their Asian neighbors that they may be motivated to pursue limited offensive capabilities, unless new incentives or new restrictions are developed within the region to make this less likely.

AUSTRALIAN SPACE ACTIVITIES

Australia has a long history of participation in space activities. Its uniquely favorable geographical location and abundant landmass has allowed Australia ready access to space technology and data from more developed space powers that have used its territory, which created a historical reluctance to undertake the expense of initiating a broad space program of its own. Since 2009, however, Australia has adopted a more active space policy, by reorganizing its domestic space bodies and planning process and seeking to develop expanded space capabilities both independently and in cooperation with key outside partners.

Space activity in Australia historically has been linked to two foreign governments: the United Kingdom and its early missile program and the United States and its civil, intelligence-related, and military space activities. These contacts led to early knowledge and expertise but also some degree of complacency. As one analyst explains regarding the problem since the 1960s: "There has been a general failure of successive governments to recognize space-related issues as being integral to Australia's national interest . . . an issue complicated by its alliance with the USA."[2] At the same time, Australia has benefited from unusual access to classified U.S. space data, putting it into a small, privileged class in terms of participation in specific areas of defense support operations.

The country's links with military space began in 1946, with the Anglo-Australian Long Range Weapons Establishment. This project eventually involved well over a thousand local support staff in and around the Woomera area supporting British missile and later other European rocket engineers, particularly from 1955 to 1967.[3] The project displaced various groups of aboriginal people and spurred protests from local church leaders and antiwar union activists.[4] Meanwhile, beginning in 1960, the United States reached a separate agreement through NASA to operate ground stations for the Mercury, Gemini, and Apollo manned missions as well as for deep-space exploration.[5] The U.S. military and intelligence community set up additional facilities over the course of the 1960s to support the Corona, Midas, Rhyolite, and Defense Support Program satellites, among others.[6]

Australia launched its first spacecraft—the *Weapons Research Establishment Satellite* (or *WRESAT*)—from its own territory in November 1967 aboard an excess (Redstone-based) Sparta rocket already in the country from a prior U.S. test program.[7] Despite its sinister name, the *WRESAT* carried only scientific instruments to measure temperature, radiation, and ozone concentrations.[8] Australia next launched a second, larger satellite— the *Oscar-5*—from Vandenberg Air Force Base in California in 1970 aboard a U.S. Thor-Delta rocket,[9] which transmitted data for forty-three days in orbit. But Australia lacked an independent rocket capability and decided not to follow up by developing its own space-launch program. When the European Space Agency (ESA) offered Australia membership in the new organization, requiring the payment of dues in exchange for participation in joint technology-development projects, Canberra turned the invitation down.[10] However, its space capabilities continued to expand with the increasing sophistication of jointly operated NASA and U.S. military facilities located on its territory.

In the mid-1980s, Australia deployed its first devoted geostationary communications satellites, the Hughes-built *Aussat* (later called *Optus*) series.[11] Its *Aussat-1* and *Aussat-2* were launched from the U.S. space shuttle, while subsequent satellites in the *Optus B, C,* and *D* series—most produced by Loral and Orbital Sciences—were boosted into orbit by commercial launch providers.

Over time, Australia has periodically been the focus of international efforts to develop commercial launch capacity from its favorable location, particularly one of the northern islands it possesses near the Equator. In the early 1990s, investors planned to develop an international facility using Russian rockets launched from Cape York, Queensland, on Australia's northern coast. But the funding never materialized, and the aboriginal community protested the project's planned use of land they consider sacred.[12] Plans turned instead in 2001 to the possible development of a major spaceport on Christmas Island after the Russian space agency offered its rocket services for the project,[13] causing the government to offer $100 million (Australian) for the development of infrastructure.[14] Again, however, the deal never reached fruition. Another proposal backed initially by Russia and then by the U.S.-based Kistler Company to develop a spaceport at Woomera for planned service to the *International Space Station* (*ISS*) and future orbital hotels was similarly stillborn. Although NASA had announced a preliminary award to support Kistler's efforts in 2006, technical problems with its spacecraft led NASA to shift its attention to SpaceX and Orbital Sciences and drop the Kistler project altogether, which then failed to generate adequate financing on its own.[15]

Despite this international interest and regular Australian *use* of space, there is still no official Australian space "program" today, at least in terms of a comprehensive, government-funded organization dedicated to space activity. One 2003 government document explained this policy by arguing: "There is no strategic, economic or social reason for the Australian Government to pursue self sufficiency in space."[16] A 2005 study observed that although Australia had the world's fourteenth-largest economy, only Mexico among the other top fifteen economies lacked a dedicated space program, and many other countries with smaller economies had much larger and more devoted space efforts.[17] Instead, Australia had a loose amalgam of academic-, private-, and government-funded space-related activities, some of which were quite sophisticated, but together lacked a sense of integration or national vision. But domestic economic factors, a fear of relative technological backwardness within the Asian region, and Australia's security needs for specific, space-derived information about its own territory have recently fostered a major push for an expanded space effort.

In 2008, the Australian Senate instructed its Standing Committee on Economics to sponsor a report on Australia's space industry and related scientific organizations to determine capabilities and consider their possible expansion. After receiving considerable expert and public input through papers and hearings, the committee's final report made a series of recommendations aimed at accelerating and consolidating Australia's space efforts. Regarding the country's past problems and official disorganization, the report summarized, "It is not good enough for Australia to be lost in space."[18] The new actions included creation of a Space Policy Unit to coordinate space initiatives within the government's Department of Innovation, Industry, Science, and Research and to establish a Space Industry Advisory Council to stimulate domestic capabilities.[19] In 2009, the Labor-led government of Prime Minister Kevin Rudd allocated an unprecedented sum of $160.5 million (Australian) for space activities, citing the country's evolving security and economic needs,[20] including new funds for competitive educational and space-related research/innovation activities.[21] Meanwhile, the Australian military's release of its 2009 Defence White Paper announced a major new focus on space activity, including the acquisition of a synthetic-aperture radar satellite for reconnaissance, enhanced satellite bandwidth for communications, and improved space situational-awareness capabilities.[22]

On the space science side, experts at the University of Western Australia have recently entered into a joint agreement with French scientists from the Tarot group to begin a joint project to expand national participation in international space research.[23] Australian experts have already developed their own capabilities in such areas as climate monitoring and weather prediction, given their unique challenges and interests.[24]

Some specialists foresee an Australian investment in a constellation of low-Earth and geostationary-orbit observation and monitoring satellites to improve the effectiveness of soil conservation and meteorology.[25] In the communications sector, Australia plans to contract with a foreign satellite producer to build two devoted broadband satellites to serve the 10 percent of Australians located in isolated inland and coastal areas who will be left out of a national fiber-optic broadband initiative scheduled to be implemented by 2018.[26]

In the military sector, Australia has continued to partner with the United States in developing its space capabilities. It has doubled the bandwidth it had previously used through an agreement with the U.S. military to share Intelsat capacity for military communications in the Pacific and Indian Ocean regions.[27] The total package involves an investment by the Australian Defence Forces of some $317 million over fifteen years. The military has also set up two space offices in Canberra's Defence headquarters (doubling its space-related staff) and begun to organize itself to train a new cadre of space-qualified specialists.[28] Finally, Australia has also agreed to participate in the U.S. Wideband Global Satcom (WGS) system by funding the sixth Boeing satellite in the planned constellation (due to be fully operational in 2013).[29] The WGS network will provide Australia with continuous wideband intelligence, reconnaissance, and communications data around the globe, with a much higher volume of information than has been available previously. The project also further enhances Australia's military space capabilities and its ties with the United States.

Australia voiced its dismay with China after the test of Beijing's antisatellite weapon in January 2007, calling in the Chinese ambassador and raising several objections. Despite Australia's participation in U.S. missile-defense activities, the Canberra government has traditionally opposed the weaponization of space and supported UN resolutions on the Prevention of an Arms Race in Outer Space. With this sometimes delicate dual role in mind, Australia is now engaging in a new project with the United States that will greatly enhance international capabilities to monitor space traffic and debris. The Australia–United States Space Situational Awareness Partnership reached in the fall of 2010 will develop new space-ranging capabilities from radar and optical stations based on Australian soil to increase greatly the alliance's catalogue of orbital objects.

Given the uncertain course of great-power relations in space and the as yet undecided fate of major space weapons systems, Australia may be able to play some role in easing U.S.-Chinese tensions and in promoting commercial and disaster monitoring cooperation among Asian space players. The Royal Australian Air Force officer (retired) space expert Brett Biddington argues that Australia is generally viewed as an "honest broker" within Asia and believes this role "could extend into the domain of space."[30] Such efforts could eventually provide it with a more important role to play in the security realm, if politics among the major powers do not become irreparably hostile in the meantime. Australia is one of only a dozen countries that have ratified the 1979 Moon Treaty, which calls for the creation of an international authority to govern and distribute profits from Moon resources, something the United States and most major spacefaring nations have long opposed. This fact, Australia's unique location, and its place (like South Korea) as a middle power in Asia could put it in good stead to serve as a bridge to the developing world in future international negotiations concerning lunar missions and possible governance issues between developed and less-developed countries.

In a major report on Australia's future direction in space, Biddington and Australia's former Director of Defence Space Roy Sach make the case that the country should be realistic about its prospects in space and pursue a balanced policy of increasing its capabilities, its infrastructure, its space-related education system, and its international relations beyond its partnership with the United States. In terms of space security, they argue that their country should "use the natural advantage of Australia's strategic geography to its best advantage in protecting and preserving the fragile environment of space."[31] Given Asia's current mistrust and trends toward space competition, such a contribution could be a very positive one in shifting regional attention to a common cause.

THE INDONESIAN SPACE PROGRAM

Given its widely dispersed islands, population, and resources, it is not surprising that Indonesia has sought to benefit from space technologies, particularly for communications and Earth observation. The country's space program is led by the National Institute for Aeronautics and Space (LAPAN, in Indonesian). Organizationally, the National Council for Aeronautics and Space (DEPANRI, in Indonesian) directs LAPAN's activities, under the chairmanship of the Indonesian president and with input from the heads of the Ministries of Foreign Affairs, Defense, State Development Planning, Trade and Industry, and State Research and Technology.[32]

After initiating a space program in 1975, Indonesia purchased its first communications satellite from the U.S. Boeing company and entered the space age with *Palapa-1A*'s successful launch in August 1976.[33] Boeing built two control stations and nine initial receiving stations for Indonesia's planned satellite network and built the *Palapa-A2*, which was launched aboard a U.S. rocket in March 1977.[34] Indonesia next orbited its third communications satellite, *Palapa-B1*, through a commercial purchase from the U.S.-based Hughes corporation in 1983 and has since built a geostationary satellite-based broadcasting system drawing largely on U.S. commercial technology. The Indonesian satellite system has provided direct television and radio programming to Indonesian homes through the Indo-Star system since 1997[35] and offers cellular phone service through a cooperative network with Thailand and the Philippines.[36]

In the field of Earth observation, Indonesia also cooperated with the United States to acquire a Landsat ground station in 1981 to begin receiving Earth resources data from space for agricultural development, forestry monitoring, and other purposes.[37] In the mid-1990s, Indonesia supplemented its Landsat data with a French SPOT ground station as well. Like many Pacific-island countries, Indonesia faces a challenge in monitoring its vast maritime domain, forcing it either to operate large numbers of surveillance ships and aircraft or to purchase significant amounts of satellite imagery from commercial companies, both at significant cost. Development of its own reconnaissance capability—even if the satellites have to be launched on foreign boosters—could reduce these costs, develop its technological infrastructure, and reduce the country's dependency on foreign technology. As a result, LAPAN is now developing its own remote-sensing satellites (including *Lapsat-1* and *Lapsat-2*).[38]

In terms of international cooperation, Indonesia has worked with the United States extensively for technology but has also signed a cooperative agreement with the Indian Space Research Organisation to host a ground station for telemetry and tracking of Indian satellites.[39] The arrangement also includes cooperation in space science and space applications. India launched the experimental LAPAN *Tubsat* microsatellite in January 2007, built in cooperation with the Technical University of Berlin.[40]

Indonesia's PT Telekomunikasi has also recently broken the prior pattern of working with U.S. corporations by signing a contract to purchase its *Telekom-3* satellite from the Russian manufacturer Reshetnev, although

Europe's Thales Alenia Space will construct the electronics package.[41] While the bulk of Indonesia's satellite network entered space aboard U.S. Delta rockets or the space shuttle, LAPAN chose a Chinese Long March 3B for its *Palapa-D* spacecraft in September 2009. The satellite itself had been built by a French-Italian Thales Alenia consortium, which had offered a somewhat more costly spacecraft than their rival bidder U.S. Space Systems Loral, but it won the contract by being able to provide an ITAR-free satellite that could be launched from the less-expensive Chinese booster.[42] Unfortunately, for Indonesia, because of a problem in the burn of the rocket's third stage, the satellite was put into a useless orbit. But by using additional fuel (at the cost of shortening the satellite's service life by five years), Thales Alenia was able to boost the spacecraft into its place in geostationary orbit above Indonesia.

Because of its consistent expenses in launching aboard foreign boosters, LAPAN is seeking to develop an indigenous space-launch capability. It has conducted multiple tests since 1987 of the single-stage, solid-fuel RX-250 sounding rocket from its Pameungpeuk launch site in West Java, reaching a near-space altitude of sixty miles.[43] In May 2008, it launched two wider-diameter RX-320 rockets. In July 2009, it upgraded its capabilities with the first launch of the three-stage RX-420,[44] although the rocket only achieved an altitude of forty-one miles.[45] LAPAN has announced plans to orbit its first small satellite by 2014. But LAPAN has considerable work to do to develop the capability of launching small satellites into low-Earth orbit, including significant budgetary hurdles. It will be some time before it will be capable of achieving geostationary orbit.

Still, the country has now set itself on a course to expand its space activities. LAPAN's budget has more than quadrupled since 2003, and Indonesia is now receiving attention from leading space powers. China sent a team of taikonauts to Indonesia in October 2010 to promote knowledge of Chinese space activities, appealing to the country's large ethnic-Chinese minority. NASA Administrator Charles Bolden visited Indonesia shortly afterward to begin discussions toward engaging this large, Muslim country in possible joint space activities. President Obama's own experiences as a child in Indonesia and his visit in November 2010 are likely to increase prospects for future joint NASA-LAPAN activities.

THE MALAYSIAN SPACE PROGRAM

Malaysia is a midsized, developing country with an unusually active space program. Indeed, it is the proverbial "mouse that roared" when it comes to its participation in international space forums and satellite activity, although both are relatively recent developments. Although Malaysia had made use of foreign remote-sensing data since the 1970s to help manage its forestry resources,[46] the country's own space program emerged after government decisions to establish a Malaysian Center for Remote Sensing in 1988 and to build a national planetarium in 1989. As the former space program head Mazlan Othman explains: "The Malaysian public had previously been quite disengaged in regards to space."[47]

But the planetarium began to galvanize greater interest in space both popularly and in the government itself. After Othman—a Ph.D. astrophysicist and leading female scientist—was named the director of the UN Office for Outer Space Affairs in Vienna in 1999, Malaysia's profile as a space leader in the developing world increased substantially. In 2002, the Malaysian prime minister, Mahathir Mohamed, recalled Othman home to create a national space agency, known in Malay as Angkasa.[48] The country's commitment to establish such a program took into consideration the domestic and international political benefits of having a space program and the "push" for Malaysia in terms of inspiring young people to enter careers in science. The space program has focused on acquiring relevant data and space technology for use in "agriculture, forestry, geology, hydrology, the environment, coastal zones, marine biology, topography, and socio-economic applications."[49]

Operationally, Malaysia has paired with the U.S. Hughes and Boeing corporations to create a robust communications network based on geostationary satellites.[50] In 1996, the Hughes-built *Measat-1* and *Measat-2* reached geostationary orbit via Ariane launches to establish a national system for data transfer and broadcasting nationwide and to points as far as eastern Australia and Vietnam. To operate the satellites, Malaysia constructed a control center on the island of Pulau Langkawi in the country's northeastern coastal region.[51] It has also undertaken active research and development to advance its domestic capability for future space activity. In 2000, Malaysia's first jointly built microsat (*TiungSat-1*), developed in cooperation with Britain's Surrey Satellite, Ltd., reached low-Earth orbit aboard a

Russian Dnepr rocket.[52] The fifty-kilogram spacecraft carried a store-and-forward communications package, a scientific experiment, and a multispectral imaging camera.[53] In order to update and expand its communications capabilities, Malaysia purchased *Measat-3* from Boeing. Launched in December 2006, it has an expected lifespan of fifteen years.[54]

In an effort to increase domestic interest and boost its regional prestige, Malaysia decided in 2005 to supplement a contract with Russia to purchase eighteen Sukhoi-30 MKM fighter aircraft with a deal via Roskomos to select and train the country's first astronaut for a mission to the *ISS*.[55] The political content of the effort was evident in the prime minister's stated rationale in ordering Space Agency Director-General Othman to undertake the program: "In my opinion, from time to time, a project comes along which can unite the nation. This is such a project."[56] This high-profile program involved picking an appropriate short list of candidates from a national competition that received 11,275 applications. A group of four candidates eventually went to Russia in 2006 for training and final selection, and this created a politically delicate challenge for the government to appease its ethnic Malay, Chinese, and Indian communities.[57] In the end, the thirty-five-year-old Malay orthopedic surgeon Dr. Sheikh Muszaphar Shukor made a successful ten-day flight to the *ISS* aboard a Russian booster in October 2007, becoming nationally famous and increasing Malaysian knowledge of and interest in space.

Malaysia has sought to develop its domestic manufacturing capability for space through cooperation with foreign partners. In July 2009, the country's *RazakSat* remote-sensing satellite—built in cooperation with South Korea—was successfully launched from a U.S. rocket.[58] The project also promoted Malaysia's space program internationally by eliciting interest in its high-resolution images, from countries in Asia, Latin America, and Africa, according to Malaysia's Minister of Science, Technology, and Innovation Dr. Maximus Ongkili.[59] Its focus is on acquiring timely ocean and meteorological data for the world's equatorial region.

After serving as director-general of the Malaysian National Space Agency from 2002 to 2007, Dr. Othman returned to her previous position as head of the UN Office for Outer Space Affairs. Thus, Malaysia continues to be represented at a high level in international space organizations, fulfilling the national goal of maximizing its limited resources through international cooperation.[60]

NORTH KOREAN SPACE ACTIVITIES

Although much attention has been focused on the attempted space launches of the Democratic People's Republic of Korea (DPRK), North Korea appears to possess no clearly thought-out plan for the development of a space industry, much less for coherent scientific, economic, or military uses of space such as reconnaissance, communications, or early warning, all of which would be useful capabilities for the isolated DPRK.

The roots of the DPRK's space program can be traced to the mid-1980s, when Kim Il Sung established a national Committee of Space Technology.[61] This body is believed still to be in charge of North Korea's limited space activities. The first real evidence of its space program, however, did not emerge until the DPRK conducted the first launch of its Taepodong I missile in August 1998 with a satellite, the *Kwangmyongsong-1* (or *Bright Star-1*), aboard. The DPRK media dutifully reported that the satellite had successfully entered low-Earth orbit and was playing recorded songs in honor of both Kim Il Sung and his son and successor Kim Jong Il. But—with the exception of a discredited Russian report—no foreign country with space-tracking facilities ever recorded these signals or provided evidence of the satellite's existence after the failed launch. Radar tracking reported in the press showed that the solid-fuel third stage and attached satellite broke up and reentered the atmosphere, ending up in the Pacific. One of the few analyses of the *Kwangmyongsong-1* satellite, conducted by Stanford University's Lewis Franklin and Nick Hansen, drawing from information revealed in the North Korean media at the time, indicates that the spacecraft was a simple .6-meter sphere with a tape recorder and six small VHF/UHF antennas.[62] The system was analog (not digital) in design and showed technology that dated to the 1950s.

A full decade passed before North Korea would attempt a space launch again, a very unusual progression for any country with any serious space intentions. The DPRK had conducted a failed test of the Taepodong II missile in July 2006, which exploded less than a minute into its flight, but without any satellite payload aboard. In the spring of 2009, North Korea announced that—despite the UN-mandated missile-test moratorium—it would be using a so-called Unha-2 rocket (believed to be a Taepodong II missile) to orbit a satellite. The DPRK government had finally gone through the steps in late 2008 and early 2009 of ratifying the 1967 Outer Space

Treaty and joining the UN Convention on the Registration of Space Objects, presumably in an effort to emphasize the "legitimacy" of its space program and peaceful intentions. It seemed no coincidence that South Korea had earlier announced plans to attempt the first launch of its KSLV-1 from its Naro space facility that June. Few observers were convinced that the purported satellite launch—which North Korea called *Kwangmyong-song-2*—on April 5, 2009, represented anything more than an illicit missile test of the Taepodong II. But the competitive politics of the Korean Peninsula suggest that the test may have purposely been a space launch attempt as well, aimed at least in large part at undercutting Seoul's upcoming launch. The Unha-2's flight, however, failed in its second stage after flying about one thousand miles past Japan and deposited its last two stages and the satellite into the Pacific Ocean, although North Korean news reports (alone) predictably reported its success.[63] All told, this is the bulk of the evidence we have about North Korea's space program. In terms of its motivations, according to the Korea analyst Daniel Pinkston: "Space launches and other spectacular scientific achievements can be exploited by Kim Jong Il and the KWP [Korean Workers Party], but they are expensive with significant opportunity costs."[64] Specifically, North Korea's apparent use of attempted space launches to continue to develop its sanctioned missile program has met with opposition even from Pyongyang's ally China and friendly neighbor Russia and has elicited further UN sanctions on the country.

Given its budgetary problems and the large amounts of funding the DPRK government spends on its military, its nuclear program, and its missile program, it is not surprising that North Korea has been unable to make any serious inroads into space. As the North Korea watcher Terence Roehrig argues, as a result of the DPRK's loss of military and financial aid after the Soviet break-up in 1991 and the rise in the comparative strength of South Korea, "Pyongyang has had to reorient its strategy from one focused on expanding its ideological influence on the South to one of regime survival."[65] Beginning in 1998, the DPRK moved to a "military first" policy to ensure that critical supplies reached the defense establishment despite hardships in the domestic economy. It also began to rely increasingly on missile sales abroad to finance its state budget, as aid dwindled and the economy continued its decline due to endemic structural problems.

Franklin and Hansen speculate that North Korea may be attempting—against all economic odds—to develop a commercial business in low-cost space-launch services. This, they argue, is the only plausible explanation for Kim Jong Il's major investment in a new western launch facility that will only have the capacity to carry out single launches, rather than the kind of barrage attacks that would be useful for the military. The satellite program, however, is perhaps more puzzling and may only be geared toward generating domestic and foreign publicity. Similar technology to the first satellite, according to their analysis, appeared in the media discussion of the second satellite system North Korea attempted to launch in 2009. These points suggest that the DPRK, in fact, has no sophisticated or devoted satellite program or serious plans to develop such an industry.

In terms of the international dimension of North Korea's space program, most of the recent attention has focused on Russia. While a number of analysts have pointed to the possible acquisition of Russian missile prototypes and periodic (and illicit) expert advice, the German missile experts Robert H. Schmucker and Markus Schiller have argued that the long-range North Korean missile "program" is mostly a charade and constitutes instead a periodic (albeit focused) effort to launch outmoded or excess Russian boosters purchased on the black market, since Russia is banned by the Missile Technology Control Regime from selling missiles to the DPRK.[66] If true, these points may help explain the extreme infrequency of DPRK missile tests and the apparent absence of a devoted test program using rocket motor test stands and "tinkering" with designs, as has occurred with all other devoted missile development programs worldwide. Schmucker and Schiller's points posit that North Korea is not likely to emerge as a major space player any time in even the intermediate future. Other analysts, however, point to collaboration with countries in the Middle East and South Asia and suggest that these states have tested North Korean missiles and related technology for them.

In particular, North Korea is believed to be using its missile technology, construction expertise, and low-cost services to provide Iran with a step-up in its space activities.[67] Since Iran cannot turn to other countries for this assistance, North Korea has stepped forward to construct a launch pad at a site near Semnan (east of Tehran), which could be used for space-launch purposes. If North Korea's missiles are being used, Tehran's greater success in launching satellites may simply be a matter of better system integration or better luck. Iran also lacks a sophisticated manufacturing capability for satellites, although, unlike North Korea, Iran has benefited from recent cooperation with both the United Kingdom and Russia in regard to space science. By contrast, perhaps because of its ideology of *juche* (self-reliance) or simply given a lack of funds and a difficulty attracting partners, North Korea is not known to have had any such contacts in the area of satellite technology. Even major Chinese educational institutions report that North Korean students no longer attend their degree programs in space-related fields.[68] This gap may simply reflect a lack of funding.

Given its missile program, North Korea could still move quickly into the category of a space-launching country. The journalist Mike Chinoy suggests that the DPRK is indeed motivated by what he sees as a "'space race' developing between Pyongyang and Seoul."[69] This point suggests that future DPRK rocket tests can be expected in an effort to steal publicity from future South Korean space accomplishments. However, it will be at least a decade before the DPRK can manufacture and operate modern satellites, and most likely much longer, given the severe gaps in its technological base. Thus, it seems that North Korea is highly unlikely to catch South Korea as the Korean Peninsula's leading space power.

PAKISTAN'S SPACE PROGRAM

After North Korea, Pakistan has the weakest space capabilities among countries that currently possess nuclear weapons. Instead of space, its government has placed a priority historically on the development of nuclear technology, solid-fuel ballistic missiles, and conventional arms. Unlike the robust space program of its neighbor and rival India, Pakistan has been unable to launch its own satellites to date and has only limited satellite production capabilities. But it does have space ambitions and, indeed, Pakistan has had an official space program for several decades. What it has lacked is the combination of prerequisites for building a well-balanced space program: adequate funding, sustained governmental attention, a strong cadre of appropriately trained scientists and engineers, and technology. In the nuclear field, the engineer A. Q. Khan stole critical designs and know-how from the Urenco enterprise in the Netherlands and used this technology to build Pakistan's nuclear bomb.[70] But sanctions from Pakistan's nuclear program and related missile development objectives have greatly limited foreign—except Chinese—willingness to share space technology

with Islamabad, as countries have been reluctant to further the development of Pakistan's nuclear delivery capabilities.

Pakistan founded the Space and Upper Atmosphere Commission (SU-
PARCO) within the Pakistan Atomic Energy Commission in 1961 to begin
research using sounding rockets, with U.S. technology and training provided at NASA's Goddard and Wallops Island facilities.[71] In June 1962,
Pakistan launched Rehbar-1—a two-stage rocket using U.S.-provided Nike
and Cajun boosters—to an altitude of 130 kilometers, followed by an iden-
tical Rehbar-2 shortly afterward.[72] France, West Germany, and the United
Kingdom also provided technology, fuel, and training for the fledgling
Pakistani program.[73] SUPARCO continued to launch British-provided Skua
sounding rockets to measure upper atmospheric winds and temperatures
in the 1970s. In 1981, the Pakistani government upgraded the SUPARCO
organization—by now with its headquarters in Karachi—into an autonomous body and established a new Space Research Council to provide an
oversight role. SUPARCO soon began to move forward with efforts to
develop its own manufacturing facilities for solid-fuel rockets as well as
related test and tracking facilities.[74] These facilities eventually developed
the Hatf-1 and Hatf-2 short-range missiles for the Pakistani military, reportedly drawing on French technology.[75]

In order to use Earth resources data provided by the U.S. Landsat program, Pakistan established a portable receiving station in Rawalpindi in
1976.[76] It later expanded this effort in 1989 by opening a permanent facility
at Rawat to enable SUPARCO to receive data from U.S. Landsat and NOAA
satellites and from the French SPOT program.[77] SUPARCO has used this
information to complete various Earth resources, urban planning, and
water resources mappings of its territory. SUPARCO now operates a number of ground stations and has opened an Institute of Space Technology in
Islamabad to help train a qualified cadre of space professionals.

Pakistan has built a small number of satellites through cooperation
with foreign partners. In 1990, the small (150-pound) *Badr-1*, built by SU-
PARCO from models developed by the United Kingdom's Surrey Satellite,
Ltd., became the country's first satellite, entering space aboard a Chinese
Long March 2E rocket.[78] Its simple store-and-forward communications
system functioned for fifteen minutes in each of four daily passes over Pakistan for a month. The country's next satellite, *Badr-2*, a somewhat more
sophisticated imaging and communications spacecraft with a radiation

detector, finally entered space in December 2001, after many delays. The spacecraft used a structure and technology designed by the UK firm Space Innovations, Ltd., although with component integration completed in Pakistan by the space agency.[79] Pakistani scientists also participated in China's *Small Multi-Mission Satellite* (*SMMS*) program, which eventually launched a spacecraft from a Chinese Long March booster in 2008. However, sources indicate that Pakistan's role was limited mainly to training in the use of downloaded *SMMS* data and images rather than any role in the development of its hardware.[80]

Looking ahead, Pakistan has entered into a memorandum of understanding with China to develop remote-sensing technology and to cooperate in the development of future APSCO initiatives in this area.[81] SUPARCO also plans to develop an indigenous remote-sensing satellite for images with a 2.5-meter resolution, to be followed by a domestically produced, all-weather, synthetic-aperture radar satellite,[82] although progress on these projects appears to be slow.

In terms of space applications, Pakistan has lagged behind many Asian countries. While it uses satellite information, it produces little of its own and has only one satellite in geostationary orbit. Ironically, given its early space program and its favorable location, with only ocean between its territory and the equator, Pakistan received an initial allotment of five slots in geostationary orbit from the International Telecommunications Union. But Islamabad's failure to use these slots with either its own satellites or with spacecraft purchased or leased from other countries caused its claims on four of the five slots to lapse. In an effort to prevent the loss of its last slot, Pakistan in 2002 leased the partially damaged, third-hand *Palapa-C1* satellite, which had been built by Hughes for Indonesia in 1996 and subsequently rented by Turkey.[83] Renamed *Paksat-1*, the satellite has experienced gap periods due to battery problems and is scheduled to be replaced. In October 2008, Pakistan contracted with China's Great Wall Industry Corporation for the purchase of a satellite (*Paksat-1R*) and associated launch services for delivery in late 2011.[84] A Chinese electronics package for the satellite will be manufactured in Pakistan with Chinese assistance, and the satellite will also fly Pakistani subsystems that will be operated for test purposes toward an eventual domestic manufacturing capability.[85]

Pakistan continues to work toward its long-sought goals of launching its own satellites and, potentially, offering launch services to other developing nations.[86] However, SUPARCO has focused primarily over the past five decades on developing solid-fuel military missiles, benefiting from Chinese and North Korean assistance.[87] It has had less success in developing reliable liquid-fuel systems for space-launch purposes, although it has announced plans to do so.[88] But developing such a program will be costly and will subtract critical resources from other, higher-placed domestic efforts, such as maintaining and developing its nuclear program, expanding its navy for Indian Ocean operations against India, and supporting antiterrorist programs to maintain control of its territory. Added to Pakistan's low literacy rates, high degree of poverty, and other economic challenges, the space program has remained a relatively low priority and may be expected to lag behind other sectors for the foreseeable future. On the other hand, some observers believe that foreign assistance from North Korea and perhaps China will accelerate Pakistan's efforts, particularly given its rivalry with India in security matters and its competition with Iran for supremacy in the Muslim world. As one Islamic Web site that tracks these issues argues, "Both Iran and Pakistan are racing towards space."[89]

Despite its relatively modest capabilities and participation as a recipient nation in training programs through the Chinese-led APSCO organization, Pakistan has also attempted to use space for political purposes. Pakistan is the host nation for the Inter-Islamic Network on Space Sciences and Technology (ISNET), formed in 1987, a membership organization that seeks to promote space education, training, and data sharing among Islamic countries.[90] To date, the activities of the ISNET group, however, have been relatively limited.

In terms of its international commitments, Pakistan is a member of all of the major UN space treaties and is among the few countries that have ratified the 1979 Moon Treaty. This point may suggest that Pakistan's signature on the treaty indicated the expected absence of any capability to conduct lunar research for some time and an effort, conceivably, to try to hamstring activities by India. To date, Pakistan does not have any known program to develop space defenses or weapons, although Islamabad has typically pledged to match India in any military capability it deploys. However, as noted above, resources are likely to continue to act as a limiting factor on Pakistan's space ambitions, at least until it solves a number of serious economic, governance, and security problems.

THE PHILIPPINES IN SPACE

Given its location and multi-island geography, the Philippines is another Southeast Asian country that stands to benefit from greater exploitation of space technology. To date, however, its space activities have been slow to develop, due to a lack of resources, trained personnel, and adequate high-level political interest. Although more than a dozen public and private organizations have some connection to space technology, only a loose, coordinating body currently organizes national space efforts, yet without budget authority or specific administrative powers. Instead, the country operates through a set of broad plans, often implemented with foreign space providers and organizations.

The history of interest in the Philippines in space-related activities can be traced to the national meteorological observatory operated in Manila from 1894 up to World War II through the efforts of Spanish and then local Jesuit clergy. While these facilities were destroyed during the Battle for Manila in February 1944,[91] the government rebuilt the observatory and, in 1954, established a separate Astronomical Observatory at the University of Philippines' main campus in Quezon City. In 1970, the Philippines acquired equipment to allow it to receive satellite images for meteorological purposes, greatly improve the accuracy of its weather forecasts, and help to improve preparedness for natural disasters. The Philippine Congress passed a bill to create a Philippine Atmospheric, Geophysical, and Astronomical Administration (PAGASA) in 1971, but the declaration of martial law by President Ferdinand Marcos delayed the formal establishment of this body until December 1972.[92] Still, it had only a minimal space component unrelated to meteorology.

In 1987, however, the Philippine government formed the National Mapping and Resource Information Authority (NAMRIA) to conduct work to assist in management of water, coastal, and agricultural resources.[93] Using U.S. Landsat and French SPOT imagery and data, NAMRIA mapped nearly all of the Philippines. A joint Philippine-Australian Remote Sensing Project in 1990 provided new incentives to both NAMRIA and PAGASA, offering resources that helped create a National Remote Sensing Center within NAMRIA.[94] Soon afterward, the Philippines created a higher oversight organization called the National Coordinating Council for Remote

Sensing and then upgraded it in 1995 to the Science and Technology Coordinating Council's Committee on Space Technology Applications (STCC-COSTA).[95] This body now serves as a national coordinating organization for space and the point of contact for cooperation with foreign space programs such as NASA and Japan's JAXA. However, critics in the Philippines believe the country would be better served by creating a single, devoted space organization rather than continuing to work via a loose coordinating body.

Before retiring in 2008, the head of PAGASA's Atmospheric, Geophysical, and Space Sciences branch, Bernardo M. Soriano, gave an interview in which he called for the establishment of a Philippine Aeronautics and Space Administration (PASA) modeled on NASA and other major space agencies.[96] While some scientists supported his proposal, certain bloggers and fiscal conservatives criticized the concept as a waste of scarce national resources in a country with a dilapidated air force and weak industrial capacity overall.[97]

The bulk of space-related activity in the Philippines centers around its use of telecommunications satellites purchased from foreign manufacturers and operated by the Mabuhay Satellite Corporation (MSC), operating out of Subic Bay.[98] Since the mid-1990s, MSC has invested in six communications satellites, the *Agila-1* and *-2*, and the *Asian Broadcast Satellite-1, -1a, -2*, and *-6*, with additional satellites planned.[99] The *Agila-2* spacecraft is a large, regionally focused satellite built by Space Systems/Loral and launched in 1997.[100] It provides television, phone, and data services to a broad area that reaches all of the major countries in Southeast Asia and South Asia. The satellite is a joint venture among Philippine, Indonesian, and Chinese companies. Some of the satellites used by the Philippines and leased out by MSC have been purchased secondhand from other spacefaring countries, such as *Koreasat-2* and *Koreasat-3*.[101]

On the diplomatic front, the Philippines is a member of all of the major space treaties, including the 1979 Moon Treaty. The Philippines has resolutely opposed activities by larger powers that have threatened to deploy weapons in space and has announced no significant plans to date for its own military uses of space, with the exception of support functions such as reconnaissance and communications. Although it will continue to use space, it may be some time before the Philippines develops significant technology of its own.

SINGAPORE'S SPACE ACTIVITIES

Singapore is a relative latecomer to Asia's space community, but it is a country that brings to bear a first-rate educational system and significant financial resources. The government has recently made space activity a high priority and is seeking to develop a domestic cadre of space engineers and experts, including in its military. As in other countries, this effort builds on prior experience in purchasing and using satellite services from other countries.

Singapore's first satellite in 1998—the *ST-1*—was built by the British-French Matra Marconi Space corporation and entered orbit aboard an Ariane rocket.[102] In order to begin building its domestic capabilities, Singapore next initiated a cooperative project between engineers from its Defence Science Organisation's National Laboratories and Nanyang Technological University to develop a small Earth-observation and communications satellite, now called *X-Sat*.[103] Singapore contracted with India's Antrix Corporation to launch the satellite in 2006, but the satellite's construction was not completed in time and delays ensued.[104] After nine years of work and a launch date that continued to slide, *X-Sat* began its planned three-year mission in 2011. The purpose of this low-Earth orbital satellite, which cost about $29 million to build, is to conduct observations of soil erosion around Singapore and to study and monitor environmental changes, with communications handled from a ground station located at Nanyang Technological University.[105]

Singapore has turned to Japan's Mitsubishi Electric Corporation to construct its next-generation communications satellite. It also has plans to continue developing its own space enterprises. As one participant in the *X-Sat* mission explained, "The main purpose of the project is to develop the capability within Singapore to design, build, test, and operate a mini-satellite bus with multi-mission support capability."[106] He notes that the range of technologies and operating systems aboard are "designed to cater for a variety of different future mission objectives."[107]

While Singapore has started late, it has some of the critical financial and technological resources to advance rapidly in space technology. It is unlikely to develop its own launch capability or a full array of space science programs, focusing instead on Earth applications, communications, and, likely, military support activities. The government of Singapore is playing an active role. Indeed, the state-led SingTel company owns the multisatellite Optus telecommunications firm, which is active throughout Australasia. Meanwhile, Singapore is sending significant numbers of its civilian and military personnel to the United States for training in engineering and space systems. This evidence suggests that Singapore is committed to carving out a clear niche in the region's space industry and in a space infrastructure to promote its economy and national security.

TAIWAN'S SPACE PROGRAM

Taiwan has long conducted space-related activities using foreign space data and jointly developed spacecraft rather than seeking complete independence. It has established some unique international partnerships to keep an eye on mainland China and to develop its own capabilities in certain space sectors. However, Taiwan's focus on other military priorities, cost concerns, possible fears of provoking China, and problems in acquiring rocket technology have left it short of deploying a workable space-launch booster to date, although it has gained expertise in satellite development. Indeed, Taiwan has focused increasing attention on space in recent years and has an ambitious government plan for expanding its domestic infrastructure and initiating a competitive space applications industry for export by 2018.

The beginning of serious space efforts in Taiwan date to the early 1990s. The island created the National Space Program Office (NSPO) in October 1991 to manage a fifteen-year government plan for space development. The NSPO began work to develop an indigenous satellite production capability with a strategy based on technology transfer via cooperative development with U.S. and European companies. In 1992, NSPO worked with the U.S. TRW corporation to build Taiwan's first communications satellite, *Formosasat-1* (also known as *Rocsat-1*).[108] Carrying a communications payload assembled by Taiwan's Microelectronics Technology and a wide-scan ocean imagery system for environmental monitoring, the satellite began operating in low-Earth orbit in January 1999 after launch from a U.S. Athena rocket in Florida.[109] Taiwan followed with *Formosasat-2*, a more sophisticated satellite launched in May 2004 to conduct Earth imaging with much-improved resolution—developed by NSPO—for observing terrestrial objects; it was equipped as well with scientific sensors.[110] As Taiwan worked to hone its Earth-observation technology, it developed an unusual cooperative arrangement with Israel that allowed it to operate Israel's *Earth Resources Observation Satellite-1* during its overflights of the Chinese mainland.[111] Taiwan also purchased higher-resolution imagery from the U.S. *Ikonos* satellite on a commercial basis, although it reportedly had to notify the U.S. government of the images it obtained.[112]

In April 2005, Taiwan's National Science Council upgraded the status of the NSPO to the National Space Organization (keeping the old NSPO acronym) in recognition of its accomplishments in moving from a role primarily as an office for technology acquisition to one responsible for technology development and system integration.[113] Besides satellite development, the NSPO has sought to expand its research in sounding rockets. By 2018, the NSPO plans to have launched ten to fifteen rockets into the low regions of space to conduct research and develop operational experience in handling instrumented payloads.[114]

Although China has frequently pressured foreign countries not to cooperate with Taiwan in space activities, the NSPO has worked with some space entities, particularly in space science. NASA has been a partner in the *Formosat/Cosmic* (or *Formosat-3*) project, a low-Earth orbit satellite providing precise measurement and navigational information for meteorology and other scientific monitoring of the Earth in cooperation with the U.S. GPS satellites.[115] Despite the relatively close relations between Russia's official Roskosmos organization and mainland China, a major Moscow-based scientific institution has worked with Taiwan since 2005 on a cooperative project aimed at studying the Earth's magnetic field and ionospheric temperatures though the development of a microsatellite known as the *Experimental Scientific-Education Micro-Satellite*, which was launched in September 2009.[116]

NSPO's planned *Formosat-5*—an independently designed and assembled imagery satellite—is two years behind schedule and is now planned for launch in the 2013–2014 timeframe aboard a U.S. Falcon 1 rocket.[117] This fifth spacecraft in the series will likely help Taiwan cross into high-resolution reconnaissance capability and the possibility of viable commercial products, although it uses some technologies provided by several

European space companies, particularly from Germany.[118] However, such cooperation is typical in the space industry and allows benefits from the international division of labor while allowing countries like Taiwan to develop technologies considered of greater national importance and/or greater difficulty to procure from abroad.

Overall, Taiwan's progress in the past decade has been relatively rapid, thanks to its high level of education and relatively advanced technological base. Taiwan's ability to continue working with the international space community has helped boost its potential. NSPO, for example, hosted the International Astronautical Federation's sixth workshop on Satellite Constellation and Formation Flying in November 2010 in Taipei. Thus far, there is no evidence of Taiwanese research into space defenses or military space capabilities beyond reconnaissance and communications, areas with civilian dual uses that make it difficult for China to raise objections or gain greater international cooperation to counter Taiwan's space activities. If political reunification with the mainland does take place sometime in the future, Taiwan's space capabilities could provide China with additional technical breath and valuable experience in working with international partners. In the meantime, Taiwan has developed core space capabilities to enable it to assist its military, advance its scientific role internationally, and create a solid basis for future commercial activities in space.

THAILAND'S SPACE PROGRAM

Thailand has had a host of reasons to become interested in the benefits of information from space, as it faces a number of problems related to the management of its natural resources, past and current security risks from its neighbors, and a domestic insurgency paid for by the cultivation, smuggling, and sale of drugs. Thanks to its relationships with the United States, China, and, more recently, Japan, Thailand has become an experienced user of space data and operator of foreign-built satellites as well as a provider of space services to other countries. It has established a solid foundation for the growth of its space industry, thanks to cooperation with more advanced space powers and its domestic efforts to institutionalize space training into its system of higher education. Since 2004, Thailand has also begun to move forward in the development of its own satellites.

During the Vietnam War, the United States began cooperating with Thailand by providing it with data from NASA's *Earth Resources Technology Satellite-1* in 1971.[119] To use this information, Thailand established the Thailand Remote Sensing Program, which eventually became a division of the National Research Council of Thailand in 1979. By 1982, the government had established the Thailand Ground Receiving Station to allow it to access information from the U.S. Landsat, the French SPOT, and a number of other Earth imaging and meteorological satellites.[120] The station was the first in Southeast Asia.[121] Thus, Thailand has had a lead on its regional neighbors in using satellite data and has since sought to establish the technical, educational, and institutional infrastructure to cement its position.

In order to build a national database of remote-sensing information derived from foreign providers, Thailand established the Geo-Informatics and Space Coordinating and Promotion Section within the Information Center of Thailand's Ministry of Science, Technology, and Environment in 1993.[122] This organizational move set the stage for the launch of Thailand's first geostationary communications satellites—*Thaicom-1*, built by the U.S.-based Hughes corporation—in December 1993.[123] A second *Thaicom-2* satellite followed the next year. Thailand established the Shinawatra Satellite Company (later renamed Thaicom) to operate its new network of spacecraft. The *Thaicom-3* communications satellite—built by France's Aerospatiale (now Thales Alenia)—entered orbit in 1997 and operated until 2006, when it was pushed into a higher parking orbit after experiencing power problems.[124] In August 2005, Arianespace launched Thailand's huge *iPStar* (or *Thaicom-4*) satellite—the largest satellite ever placed in geostationary orbit, built by the U.S.-based Space Systems/Loral—which offers broadband services for the broader Asian-Pacific region.[125] Australian service providers, for example, have become major users of *iPStar* for broadband communications in the Ku-band.[126] The *iPStar* satellite marks the expansion of Thailand's space enterprise.

In November 2000, Thailand upgraded its developing space organization into the Geo-Informatics and Space Technology Development Agency (GISTDA) to serve as the coordinator of its space activities and to begin organizing the country's move from being a consumer of space-related data to building satellites and becoming a provider of such information.[127] In 2004, GISTDA signed a contract with EADS-Astrium to co-develop the *Thailand Earth Observation Satellite* (*Theos*) for remote sensing. The satellite successfully reached low-Earth orbit in October 2008 after its launch aboard a Russian Dnepr rocket. The *Theos* spacecraft features a two-meter, panchromatic optical imager and a wide-swath, fifteen-meter, multispectral imager to provide information useful for a range of Earth-observation missions, including land use, coastal management, agricultural monitoring, and cartography.[128] GISTDA plans to market *Theos* data internationally and establish ground stations in interested countries.[129] As part of the *Theos* contract, seventy Thai scientists and engineers received training from EADS-Astrium to assist in the operation of the spacecraft, which has a planned service life of seven years.[130]

Although Thailand has cooperated extensively with France and Japan, the government also has decided to participate in training and joint research programs with China and its AP-MCSTA organization. In 1992, Thailand participated in the first AP-MCSTA meeting in Beijing and has since taken part in all of its major events.[131] It was a participant in the *SMMS* project (mentioned earlier) and has plans for further cooperation with this group of countries in order to enhance Thailand's space education and research opportunities. Indeed, Thailand has become a formal member of the Chinese-led APSCO, indicating a notable willingness to tie itself closely to China's space program. This move is consistent with recent trends in Thailand's foreign policy.

Thailand's progress as a spacefaring nation has not proceeded without some glitches, given the difficult political situation in the country over the past decade. In 2010, the government forced Thaicom (which operates Thailand's geostationary satellites) to shut down broadcasts from the anti-government People Channel Television (PCT) company.[132] Thaicom had to resort to electronic jamming of PCT's signals in order to disrupt local reception of the broadcasts.

Thailand has faced recent challenges in maintaining its early lead among Southeast Asian countries, particularly as highly motivated and organized countries like Malaysia and Vietnam have made great strides in recent years. However, Thailand has been successful in developing considerable expertise through training programs organized by China's AP-MSCTA and APSCO and Japan's APRSAF as well as joint commercial ventures with European space companies. Within Thailand, Naresuan University has initiated a master's degree program in space technology and geoinformatics, focusing on both the manufacture of space hardware and the use of space-derived data.[133] The program has benefited from expertise provided by a regional GISTDA center established earlier at the university.

Thailand served as the host of the sixteenth annual meeting of the APRSAF organization in January 2010, where it was able to showcase its *Theos* ground stations and progress in space operations, data development, and education. Although not the first international space conference held in Bangkok, the APRSAF-16 session, involving some three hundred participants, marked a "coming out" party of sorts for Thailand's space program and its movement toward more robust national capabilities. In the future, GISTDA is hoping to use assistance from APRSAF to fund the construction of facilities for the manufacture of Thai-produced microsatellites and to cooperate with Japan and other countries in building small satellites as well.[134]

THE VIETNAMESE SPACE PROGRAM

Vietnam has a surprisingly long but discontinuous history as a spacefaring nation. In 1980, its first (and only) cosmonaut Pham Tuan flew aboard a Soviet rocket to the *Salyut 6* space station. But, given the country's internal economic problems and relatively low level of technical development, few space-related activities followed this isolated accomplishment in the subsequent decade. As part of its ambitious economic reforms since the 1990s to open up to the outside world, however, Vietnam initiated a significant space program involving extensive cooperation with a range of countries, mainly in the capitalist world. These efforts began largely after 1995, when Vietnam initiated a project to purchase its first communications satellite as a means of modernizing its domestic telecommunications industry.

Among other contacts, Vietnam has benefited particularly from training and Official Development Assistance (ODA) from Japan and its APRSAF organization. Other cooperative activities have involved the European Space Agency and companies and universities in the United States, South Korea, and Malaysia.[135] In addition, Vietnam has participated in forums sponsored by the UN Office of Outer Space Affairs.[136] Japan is scheduled to launch Vietnam's *Pico Dragon* cubesat—a two-kilogram remote-sensing spacecraft—in late 2011, thanks to a cooperative program involving technical assistance from Japan's JAXA.[137] A follow-up project is already planned to build a larger satellite. Yet Vietnam has not relied only on Japan and is also developing its own capabilities through both purchases and other assistance programs. Overall, Vietnam hopes eventually to develop its own satellite industry as a means of stimulating national technological advancement.

Vietnam's first geostationary communication satellite, *Vinasat-1*, was built by the U.S. Lockheed Martin company and launched successfully aboard a French Ariane booster in 2008. Vietnam currently operates two ground stations to receive data from *Vinasat-1* and is building an additional facility.[138] *Vinasat-1* has a minimum estimated service life of fifteen years.[139] Although the satellite project was valued at $180 million, Vietnam fulfilled its goal of ending its dependence on a $15 million yearly lease for satellite services from Thailand.[140] A larger and more sophisticated *Vinasat-2* geostationary satellite is now under contract with Lockheed Martin, with a launch date scheduled for 2012 from the Arianespace facility in French Guiana.[141] For the Vietnamese people, however, the significance of *Vinasat-1* is not limited to its technical or commercial value. Vietnamese scientists emphasize its social and political significance for such a postcolonial, communist, and previously isolated country, calling it an event that "will help raise Vietnam's image in the international arena" and describing it as "a memorable milestone for Vietnam and its integration into the world economy."[142]

Vietnam is also working on a project called the *Vietnam National Resources, Environment, and Disaster Monitoring Small Satellite-1* (or *VNREDSAT-1*). The aim of this effort is to build an indigenous spacecraft for national resources and disaster monitoring for planned launch into an elliptical orbit over Vietnam in 2012.[143] Vietnam has suffered four hundred deaths and $1.5 billion in losses on average each year from typhoons, flash floods, and soil erosion.[144] The project is a highly ambitious one that is benefiting from French and Belgian ODA funds and technical training from France in satellite technology and space operations.[145] Notably, Vietnam has not cooperated with its fellow communist country China in its efforts to enter the international space community, reflecting the continuing role of strategic concerns in regard to its large neighbor, which carried out a bloody, month-long invasion of Vietnam in 1979.

In order to manage and develop its space assets, Vietnam founded a Space Technology Institute (STI) in 2006. By 2010, STI had a growing staff of fifty-seven and has plans to expand this number by outreach to technical universities.[146] The government's goal thus far has been to begin to apply space technology to the communications, meteorology, agricultural, navigation, and environmental monitoring fields, largely through the use of foreign satellites and related data.[147] During the 2011 to 2020 period, however, STI is charged with mastering the operation and production of small satellites, Earth stations, and even launch facilities in order to give Vietnam an autonomous space capability.[148] New applications for the use of GPS technology for coastal construction projects and marine management is also planned, including installation of GPS receivers on 1,500 fishing boats to assist in weather monitoring and rescue operations.[149] Vietnam plans to open the Hoa Lac National Space Center near Hanoi by 2017, which it is building with $400 million in ODA funds provided by Japan.[150] The facility will serve as the hub for the country's small satellite manufacturing activities.

Starting from a relatively limited base, Vietnam's telecommunications industry has surged in recent years, with an annual growth rate of over 20 percent and revenues of over $6 billion.[151] *Vinasat-1* is anticipated to allow the country to achieve 100 percent reach with telephone and television services into the Vietnamese countryside. Vietnam has claimed four geostationary slots with the International Telecommunications Union, although it is currently occupying only one.[152] *Vinasat-2* will take over a second slot. Given its wider coverage, Vietnam hopes to rent transponders on *Vinasat-2* to potential users in Laos, Cambodia, Thailand, and possibly Burma.[153] But Vietnam has only until 2012 to occupy its other two slots.[154] Given the limits of its resources, Vietnam may choose to lease these slots to other countries until it is able to occupy them with Vietnamese-owned spacecraft.

As with many developing countries, space technology is seen as a means of jumpstarting both science education and national economic development. STI Associate Dean Pham Anh Tuan explains this process and Vietnam's vision by saying: "With basic knowledge like mathematics, physics, mechanics and IT, space technology engineers can seek jobs in not only the traditional space technology industry but in other fields, such as automobile, petrochemistry, oceanography, research and development centres, government agencies, universities, etc."[155] For Pham, the key challenge ahead for his nation is to "popularise knowledge about space technology and its great advantages to social and economic development, in sustainable development[,] and [the] growth of Vietnam."[156] Looking ahead, Vietnam plans to establish a governmental Vietnam Space Committee to coordinate national space initiatives and its international cooperative efforts.[157]

CONCLUSION

A substantial number of Asian countries are carrying out active space efforts. As late developers, they have been able to start out by using foreign-provided space data and then purchasing technology for devoted national purposes, rather than having to go to the expense of developing their own space infrastructures immediately. Given the rapid expansion of commercial space services in the past twenty years, such strategies have allowed them to make rapid progress toward the effective use of space. By relying on foreign technologies, they have been able to enjoy the benefits of globalization while they work to establish their own space-related domestic technological bases and the educational systems to support them. Despite their current reliance on outside technology, however, it is notable that almost all of these countries have sought to develop their own national satellite manufacturing sectors, particularly for remote sensing and communications. Fewer have gone as far as seeking an independent launch capability. Given the high costs associated with rocket technology and the wide availability of launch services on the international market, such strategies may represent the most cost-effective options for countries seeking to gain value from space-derived information without having to compete with the major space powers or accept the major start-up costs associated with trying to catch up with Asia's leaders. But few are shying away from space.

Indeed, space activity is now seen by many countries across Asia as not only an essential part of the modernization of their economies but as a critical social tool for stimulating their nations' educational institutions and raising the interest of the next generation in science, engineering, and information technology. There is also a clear connection between space and success in overcoming obstacles in land use, coastal management, disaster prevention, agricultural production, urban planning, and, from a broader perspective, national governance. Specifically, space technology allows governments to reach their populations more effectively, provide information beneficial to their economic prospects, and allow them to use modern services (such as precision navigation) provided by global utilities. Thus, are these countries truly engaged in a space race? Yes, but it is clearly a slower one than has appeared among the largest and most capable Asian nations. These countries will lack the wherewithal and infrastructure to conduct a full range of space activities for at least a decade or more, particularly in such high-prestige areas as independent space science missions or human spaceflight. Yet a number of them are already developing their space capabilities with a clear eye on their neighbors and are using space to keep informed about their military activities. Thus, one cannot neglect their participation in Asia's space competition and in the increasing use of space globally: forces that have made space a very different and more complicated environment than it was during the cold war. With these challenges in mind, the final chapter of this book examines the impact of these changes on prospects for successful regional management of space activity and, in particular, for conflict prevention.

CHAPTER SEVEN

ASIA'S SPACE RACE

Implications for Regional and Global Policy

There is a space race going on in Asia, but its outcome—peaceful competition or military confrontation—is still uncertain. Fortunately, the dynamics of space activity in Asia are not solely hostile, nor are they intrinsically military related. Thus, there are still reasonable prospects for avoiding negative outcomes in space. Yet as countries in other parts of the world are moving toward closer cooperation in space, Asia is at risk of lurching backward, motivated by historical mistrust and animosities and hindered by poor communications on security matters. Military space expenditures are rising rapidly among the leading Asian space powers, and there is resistance to the idea of country-to-country or regionwide negotiations on confidence-building measures. Preference instead is given to general statements in support of space arms control and "peaceful uses," while some national militaries have begun to develop significant capabilities and to consider different forms of space defenses and even offensive systems. India, for example, has long held itself as an example of a country supportive of peaceful uses of space and the prevention of an arms race, but even it is now moving toward space weapons capabilities. Such trends may portend a hardening of regional perspectives and increasing difficulty in moving forward with space security options.

The problem at present has not been the absence of a common desire for stability in space but instead inertia and inaction on the space security front by the leading Asian space actors and the United States. There has been much posturing but few practical steps toward creating new mechanisms for preventing space conflict and facilitating cooperation, creating a situation where countries are opting for traditional nationalist reactions to emerging threats rather than crafting new, collective security solutions. Yet, as the Singapore-based scholar Mely Caballero-Anthony argues, the path forward is as clear in space as it is in other fields of nontraditional security (NTS): "Declarations of intentions and soft commitment have to give way to more common action in solving common problems, more binding commitments, and more credible enforcement of regional agreements or modalities adopted to address different types of NTS challenges."[1]

From the realm of political economy, many of the concepts of late economic development discussed in chapter 1 predict the likelihood of problems and the emergence of competitive and even aggressive forms of nationalism. The cold war–era theorist Alexander Gerschenkron saw a direct line from the kind of state-led efforts in technological innovation seen in Asia to increasing authoritarianism and militarization.[2] Some modern writers on space have developed the same conclusion about the likely propensity of at least some of the leading space powers in Asia to end up in conflict. The potential for a bad end to Asia's space race is real.

But other authors have rejected the inevitability of any such linear connection. Certainly, much has changed since the cold war. The establishment of postwar democracies in Japan, India, South Korea, the Philippines, and elsewhere, as well as significant reforms in many others, has changed the political direction of most of developed Asia. Those countries with the most authoritarian regimes (such as Burma and North Korea) tend to be the least developed technologically. China has evolved politically and especially economically since the 1970s and continues to do so. While the Chinese Communist Party (CCP) and its Vietnamese counterpart remain firmly in power, for example, they both have very different objectives and policies than those of their initial revolutionary leaderships. Indeed, economic factors and the drive toward both modernization and integration into the world economy have clearly played significant parts in changes throughout Asia.

Given these concerns, countervailing cooperative or at least peaceful trends are likely to affect Asian space decision making and might, as many authors have argued, reduce the likelihood of future space conflict and the intensity of the space race, if not necessarily encouraging an actual rapprochement among current rivals. As Muthiah Alagappa argues, "contrary to the widespread assertion that Asia is a dangerous place . . . international political interaction among Asian states, is for the most part, rule-governed."[3] The question is: will the exceptions to these emerging norms—such as military space activity—be brought into a reliable management framework fast enough to prevent the outbreak of conflict? The record on this score is not altogether encouraging. As Bates Gill and Michael Green observe, the "history of multilateralism within the Asia-Pacific region demonstrates continued underinstitutionalization compared with Europe and the Western Hemisphere, repeated moves and countermoves by states to initiate or manipulate multilateral institutions for balance-of-power purposes, and a heavy reliance on U.S. alliance networks to provide public goods."[4] Progress takes time, particularly in a region with so much historical conflict and so many cultural differences. As noted earlier, Europe has set a positive example. It has had access to potential space weapons technology for at least three decades but to date has rejected it in favor of cooperative, civilian space activity. In this chapter, we will examine the implications of these contradictory trends as well as what might be done to set the region on a more cooperative course through new policy initiatives. The analysis begins at the regional level, dividing the discussion into the relevant civil, commercial, and military dimensions. It then examines these same areas in regard to international risks. Finally, it considers how these risks might be better managed by policymakers and how future conflict might thereby be prevented. The chapter concludes by arguing that while the heavy lifting to achieve cooperation will have to be done by Asian leaders themselves, outside countries and organizations could play a useful role by highlighting opportunities and seeking to alter incentives by providing a more favorable international context.

REGIONAL IMPLICATIONS OF ASIA'S SPACE RACE

At the outset of any discussion on the future of regional relations in space, it is worth reiterating that space activity is a *subset* of broader political, economic, and strategic relations within Asia. However, unlike trade or even military-to-military contacts, space has frequently been sequestered as an area considered too sensitive to discuss and where cooperation remains highly restricted. Ironically, these trends go against the evidence of increasing openness in the region's economies, expanding free-trade pacts, and heavy investment in one another's industries and labor markets. In this regard, space remains a bastion of nationalism, and accomplishments in space continue to be portrayed as existential victories in the history of each nation and its struggle for both the accoutrements of modernity and relative advantages over regional rivals. For these reasons, closer regional cooperation has eluded Asia in space to date, in contrast to Europe, where space accomplishments are identified as "shared" and space-related secrets are few.

In this context, several space-related questions demand attention from individual Asian countries and by the region as a whole: What are the implications of the current trajectories of national space policies? What areas might become sources of conflict? Are the national investments being made in particular areas of space capability sustainable? In what ways is space likely to be affected by the forces of globalization? Are military aspects of national space policies likely to conflict with civil and commercial aims? And, finally, where are unilateral national policies in regard to space security likely to bump up against regional and international interests for strengthened collective security in space? To address these questions and to seek answers, it is necessary to break down Asian space dynamics into the three major areas of space activity—civil, commercial, and military—and to review the status of space competition as well as where and how gaps in cooperative efforts may need to be addressed.

Civil Space Activities

Civil space activities include space science, exploration, space applications, and human spaceflight. These have typically been areas of space activity that have been funded by national governments, since these activities tend not to generate commercially marketable goods and services. Instead, their main "outputs" are knowledge, know-how, and national prestige, such as those that came from the U.S. Apollo program and the 1969 Moon landing. In Asia, there have been a number of smaller equivalents: the Chinese *Shenzhou* and *Chang'e* missions, the Indian *Chandrayaan-1* lunar orbiter, and the Japanese *Kaguya* mission and the *Kibo* space module for the

International Space Station (ISS). None of these activities generated a "profit," but each of them raised national pride and caused regional neighbors to take notice. Although the *Chandrayaan* also carried international pay-loads, these missions aimed primarily at furthering national interests, par-ticularly since most of their scientific accomplishments were redundant to those already made by the Russians and Americans in prior decades.[5] Fu-ture such nationally oriented projects are being planned by all of Asia's major spacefaring nations. This is not necessarily negative, but continuing to block intra-Asian cooperation may be shortsighted.

The rivalry brewing in human spaceflight threatens to become the most troublesome. China has taken the lead in independent human spaceflight, but Japan and India are potential competitors. All told, Japanese astro-nauts have more space experience than China's taikonauts, albeit on Rus-sian and U.S. spacecraft and the *ISS.* With its new *HTV* shuttling cargo to the *ISS,* Japan may also decide to human-rate its H-IIB launch vehicle to give it independent human access to low-Earth orbit. Given its strategic rivalry with China, India is likely the most highly motivated challenger to China in independent human spaceflight. It has plans to human-rate its GSLV rocket and to orbit a two-person capsule by the end of the current decade. India is also working with Russia on its program, which might eventually involve flights to the *ISS.* Whether this competitive approach to space makes sense or is sustainable for India remains to be seen.

Evidence of cooperation in environmental monitoring and disaster pre-paredness has thus far provided one of the few areas where there seems to be some convergence of interests and a willingness to cooperate. Since the 2004 Southeast Asian tsunami and the 2008 Sichuan earthquake, states have begun to organize themselves for data sharing and mutual assistance. If these systems can be developed into broader transparency measures for the region, there could be gains in substantive cooperation and political benefit in diffusing tensions. K. K. Nair reminds us: "It is . . . an established fact today that outer space has enormous scope for bettering human wel-fare as well as fostering civil development; Asia is in dire need of both."[6] But he cautions that Asia cannot continue on its current path toward con-flict and "watch significant gains being frittered away on account of per-ceived insecurities."[7] If Asian countries can learn to work together using space-derived data, they will be more successful in combating the common threats posed to the region by natural disasters and the effects of global warming on sea levels, riverbanks, crops, desertification, and weather patterns. The question, therefore, is whether the fears and energies that drive space rivalries can be turned into a common direction.

Some experts believe that turning the focus in the other direction—that is, away from Earth and instead toward joint efforts in exploring space— could be an easier route and help avoid the problems of politics on the ground. As the long-time NASA observer Roger Launius argues: "All the promise held out for space flight in gaining scientific knowledge, advancing technology, and creating a hopeful future through exploration of the solar system may well pale in comparison to the very real possibility of enhancing cross-civilization relations through this one act of working together to take an enormous challenge."[8] Launius observes further on the value of cooperation that "spending a larger share of the public treasury for space exploration is eminently better than spending it for weapons of destruction."[9]

One potentially positive development toward breaking the ice in civil space between China and Japan and their rival regional space organizations was the visit by JAXA Vice President Yukihide Hayashi to APSCO's headquarters in Beijing in mid-October 2009. The two sides indicated a willingness to explore future cooperation, and Secretary General Zhang Wei pledged to send an APSCO observer to future APRSAF meetings.[10] However, APSCO's own meetings have not been opened to nonmembers, so questions remain about China's actual openness to substantive cooperation with APRSAF. Handberg and Li cogently observe: "China's present status with regard to most of its current partners is clearly that of leader— a position easily maintained when the relevant partners are significantly weaker. Adding Japan to APSCO would change the dynamics since Japan if so motivated can compete with China with regard to space technologies and their applications."[11] If China believes its status is threatened, it may reject such cooperation.

Commercial Space Activities

Commercial space developments in Asia today present perhaps the most hopeful signs of movement toward a new model of space activity. As elsewhere in the world, Asian space operators are seeking to reach across national boundaries in providing services and acquiring technologies. Commercial space activity has become increasingly international since the cold war, although rigid export controls adopted by the United States in 1999 have left U.S. industry hamstrung in competing in much of Asia. Asia's space enterprises, although often remaining in state hands or under considerable state influence in terms of their activities, have become increasingly active in marketing their services internationally, including launchers (China, India, Japan), satellites (China, India, Japan, and South Korea), remote-sensing services (several countries), broadcast capacity (many countries), and direct-to-home services (many countries). To date, communications services have been the most lucrative, but there is a growing market for higher-margin sales in satellite production and, at least prospectively, in launch services. As the Indian space official Narayana Moorthi argues, "While it is possible to be totally self-reliant, it may not be prudent or necessary to develop systems which are available on the global market."[12] For developing countries, however, the problem is that "the assurance of such a supply of space applications does not exist,"[13] thus fueling efforts to develop independent national capabilities rather than an international division of labor.

At the same time, many Asian countries have attempted to keep their markets closed to competition from foreign providers to protect their domestic space industries. In India, for example, some progress has been made of late, but obstacles still remain. The Telecom Regulatory Authority of India, for example, approved rules in June 2010 allowing for the first time major foreign ownership (up to 74 percent) in direct satellite television companies operating in the country. However, the U.S. Satellite Industry Association has complained that such companies are still blocked from using foreign satellites to provide that content to the Indian market. Instead, they must use India's own Insat system or contract through the Indian Space Research Organisation (ISRO) to lease space on a foreign satellite selected by ISRO, which pockets a fee. An editorial in the industry weekly *Space News* commented on this problem: "The main beneficiary of this protectionist regulatory scheme, is, of course, ISRO."[14] For this reason, there are still hurdles to cross before anything resembling an open market exists in much of Asian space activity.

A related dilemma for even some of Asia's major space powers is that they have built space-launch vehicles and satellites at great cost for domestic reasons but cannot produce them cheaply enough to sell on the international market. Japan, for example, has long sought clients for its H-II rockets but only very recently has been able to attract even a few foreign contracts, given its noncompetitive pricing. With their lower labor and technology costs, China and India have been comparatively more successful at entering the international marketplace for launch services. At the same time, if Japan had decided to save money and opt for foreign launchers, it would have been dependent on the space technology of others, something that Japan and other leading Asian powers have deemed unacceptable for political, economic, and security reasons. Establishing a regional division of labor would, however, seem to be a logical option for Asia, given the diversity of services available at relatively lower prices than those offered in the United States, Europe, and even Russia. Specialization is likely to become a necessity for smaller space actors trying to enter the market, like South Korea, which is unlikely to be able to develop a full range of space services within the limits of its own national budget. But closed markets will make this very difficult. Partnering with other countries for space barter deals (satellites for launch services, human spaceflight access for communications services) could represent one means of overcoming these limits. Joint development of spacecraft, as occurs frequently in Europe, is another possible route.

Military Space Activities

Although much of the public rhetoric among Asian space actors downplays notions of a space race, the behavior of the major countries in space belies such restraint and cooperative sentiments. With very little digging, the sharp competitive motivations of space activity can be seen, as well as space's role in larger tactical and strategic struggles with key rivals for power and influence within Asia. Although China does not yet have a broad-based military capability in space and lags considerably behind both Russia and the United States, observers worry that its continued rise could well stimulate regional rivals, particularly if China is perceived as developing capabilities that might further shift the overall military balance in its favor. As India's Nair writes: "Apart from the military force balance in both conventional and nuclear terms which is overwhelmingly in China's favor, the harnessing of military space capabilities would allow it to enhance its capabilities manifold."[15] He warns, "China's military modernization programme . . . is overwhelmingly focused on enhancing its aerospace capabilities." In terms of purely self-interested policies, increases in military space capability will help major Asian countries project force outside their borders. As James Mackey observes, "The dual use of civilian and military rockets being developed and placed into operation by several countries (e.g., Israel, Iran, North Korea, and India) opens the door to rapid growth in the potential weaponization of space."[16]

Such policies are also likely to lead to conflict with or at least challenges to their neighbors. Mackey concludes, "As China's dependence on satellites grows, so will its vulnerability, forcing senior leaders to pursue a more robust ASAT capability or abandon such efforts entirely."[17] This latter option may be the most important objective toward securing safe access to space for all regional players. Unfortunately, efforts to reduce U.S.-Chinese military tensions have proven difficult in recent years, given the periodic flareup of tensions over such issues as U.S. arms sales to Taiwan, which led the Chinese military to reject U.S. efforts at talks, including on space security matters. As Secretary of Defense Robert M. Gates argued in June 2010: "Nearly all of the aspects of the relationship between the United States and China are moving forward in a positive direction, with the sole exception of the military-to-military relationship."[18] While his comment indicates the many areas of common interest between the two sides and their flourishing economic relationship, the absence of stable military relations marked a critical exception. After his trip to China in January 2011, Gates expressed concern about the Beijing military leadership's unwillingness to agree to an in-depth strategic dialogue and about Hu Jintao's apparent ignorance of the PLA's recent test of a stealth fighter aircraft.[19] As with the Chinese 2007 ASAT incident, these comments heightened Western concerns about a possible "disconnect" in civilian control over the PLA or, at least, in civilian oversight. Nevertheless, Gates rejected the notion that China represents an "inevitable strategic adversary" of the United States and welcomed the PLA's decision later in the year to accept a limited strategic dialogue. The question is whether the two sides will be able to establish a stable, regularized interchange and agree to new norms for conflict prevention. The same goes for Asia's hostile dyadic relationships, where such contacts are proceeding even more slowly.

GLOBAL IMPLICATIONS OF ASIA'S SPACE RACE

While regional tensions in twenty-first-century Asian space activity pose one set of concerns, the possibility for spillover onto the global scale are a second and closely related problem. Given orbital physics, any future conflicts involving Asian space actors will likely cause problems for other space actors as well. Similarly, ground-based weapons located in Asia for use against space—such as China's direct-ascent ASAT system—will pose a threat to any spacecraft as it flies over, at least for those in the low-Earth orbital band. Moreover, such shifts in regional capabilities could affect the global power balance. Kevin Pollpeter argues, "China's efforts to develop its space program to transform itself into an economically and technologically powerful country may also come at the expense of U.S. leadership in both absolute and relative terms."[20] Relative decline by Russia and the European Space Agency—and perhaps the United States—is also likely as China and other Asian space powers increase their capabilities, as has happened in many other industries, such as shipbuilding, automobiles, and electronics. These factors, however, need not be disruptive if other actors adapt successfully to the new competitive environment and find ways of using Asia's new strengths to their benefit, as Asian consumers seek new space products and as their entrepreneurs begin to invest overseas in high-tech joint ventures. But transitions are always difficult and could be met with resistance from more developed space powers eager to maintain their advantages and by rising space programs focused on becoming full military members of the space "club."

Civil Space Activities

The trend in civil space within Asia seems to be repeating some of the negative and competitive trends of the cold war between the United States and the Soviet Union. This can be seen in decisions by Asian capitals to undertake high-prestige missions to asteroids and the Moon and to move toward autonomous—rather than joint—capabilities to conduct human spaceflight. Unfortunately, these Apollo-type projects represent highly competitive efforts and may risk further exacerbating international tensions in space. As the space expert John Logsdon notes regarding the cold war's race in human spaceflight: "Although the US lunar landing program did succeed in meeting its political objectives, Apollo should not serve as a model for future human exploration."[21] Logsdon cites budgetary concerns, a decision to put science on the back burner, the failure to have a true strategy for the Moon's settlement, and the role of the space race in heightening, not dampening, political tensions. He notes that the overriding goal should not be national achievements or relative gains among countries but instead shared objectives such as "making humanity a multi-planet species."[22] The challenge is to decide how this might be accomplished within a workable, cooperative framework.

India's Moorthi argues in regard to the high costs of space activity that "Whether accepted now or not, the future technological advances in areas like reusable launch vehicles and complex interplanetary missions may require the integrated efforts of many nations."[23] He notes the "structural transformation" that has led to increasing international contacts in recent years within the space industry and observes, "international cooperation in certain areas is emerging as a necessity, side by side with competition."[24] Will these factors promote regional cooperation and a more rational division of labor in Asian space activity in the future, or will they continue to support duplicative national programs out of mistrust and a concern over prestige?

A related question is how long Asian countries will pursue separate human spaceflight efforts. Could major Asian spacefaring nations combine their efforts through the *ISS* or other projects? China remains an outlier, currently, largely out of opposition to its inclusion by certain space station members, most particularly the United States and Japan. On the other hand, a sincere invitation and a promised central role in future joint missions by the United States, Russia, Japan, and the European Space Agency might well cause a shift in these dynamics, given the tremendous costs for China in building its own systems and the prestige that could be accorded by achieving its long-sought acceptance into the club of *ISS* partnership members. Handberg and Li argue: "For China, the decision as to how to work with others will be crucial for keeping its space program on track and within the budget the leadership desires."[25] The same may be true of India, which too has not been an *ISS* partner to date, although more because of a lack of financing and experience than any specific restrictions.

But such transnational approaches to space exploration have critics. The Heritage Foundation's Dean Cheng argues regarding potential U.S.-Chinese missions: "Such cooperation has far more potential cost than benefit."[26] Eric Sterner of the George C. Marshall Institute makes a similar case in arguing that efforts to change China's space policy through cooperation are bound to fail. He warns: "Given conflicting interests, Americans will not be able to 'steer' Beijing's space behavior through the promise of a close partnership any more than King Canute could order the tides to stop."[27]

Congressional reaction to NASA Administrator Bolden's trip to China in October 2010 made clear that the issue of human spaceflight coopera-tion remains politically contentious in Washington. U.S. Representative John Culberson, a Republican from Texas, exemplified conservative oppo-sition by stating in a letter to President Obama: "I have grave concerns about the nature and goals of China's space program and strongly oppose any cooperation between NASA and [the China National Space Adminis-tration's] human space flight program without Congressional authoriza-tion."[28] However, as supporters have noted, civil space cooperation took place with the Soviet Union exactly at a time when it was a serious peer competitor in order to defuse tensions and learn more about its technolo-gies. The same arguments could be made today in support of *ISS* participa-tion by China.

Two former U.S. government officials have argued that one of the best ways to ensure continued advancement of U.S. science and to learn more about China's direction and capabilities is to promote—not prohibit—ex-changes in basic science.[29] In space, this could mean lifting current ITAR restrictions on particular areas of space research (such as life sciences, space weather, and planetary geology) on U.S. universities as a means of fostering mutually beneficial cooperation. As Bruce Bade and William Berry argue, important side benefits of such efforts would be confidence building and a reduction of the overly broad secrecy blanket that has affected all of U.S. space activity, to the detriment of U.S. science and security.

Gregory Kulacki and Jeffrey Lewis suggest that the best way that U.S. policymakers can learn about Chinese intentions "is not for outside ana-lysts to read 'tea leaves' scattered in individual Chinese actions, public statements, or academic publications" but rather to engage officials in Bei-jing in a "sustained and broad-based discussion of the civilian and military uses of outer space."[30] The agreed-upon statement on space that ended up being included in the November 2009 Beijing summit documents outlined perhaps the beginning of such a process. However, progress has still been very limited.

Commercial Space Activities

While many Asian states followed Russia and the United States by starting space programs as state-run activities, market pressures have now caused the two original spacefaring nations to move increasingly toward commercialization and integration. Given the trends in China's economy and the leadership's desire to use space for not only political and strategic but also for economic benefit, these forces are likely to affect China's space program as well, despite the currently large role that its military plays.

The most likely candidate to play an expanded role, at least in the short run, is the Great Wall Industry Corporation (GWIC). After a number of years when GWIC was unable to market its launchers to international customers, given U.S. export controls on foreign satellites that contain U.S. technology, the development of ITAR-free satellites by Western countries seeking cheaper access to space has allowed GWIC to resume successful commercial operations. Yet Handberg and Li argue that GWIC "presents the image of privatization to the world but the reality is total government control rather than market judgments."[31] China's failure to open up opportunities for smaller, private firms and growing costs for government-run space activities may become an increasing burden. One exception to this rule has been the Hong Kong–based firm APT Satellite Holdings, which has remained in private hands and now generates 74 percent of its revenues from customers based outside of China, including markets in Australia, Indonesia, and Singapore, as well as parts of Africa and Europe.[32] Even in the midst of the global economic downturn in 2009, ATP managed to increase revenues by 43 percent.[33] Such high performance may challenge the Chinese government eventually to scale back its total control over the rest of the country's space industry.

In the United States, private firms are driving industry profits from such products as direct satellite television, mobile navigational services, and other communication-related technologies. New U.S. rocket companies— such as SpaceX, Scaled Composites, and Bigelow Aerospace—are also moving into the area of low-cost launch services or private human space-flight into low-Earth orbit. Despite the role of the private sector, however, U.S. commercial contacts with foreign entities have been tightly restricted by governmental controls since 1999. But after a decade of such overly protective controls, which have harmed U.S. space cooperation even with allies, many experts, military officials, and industry personnel have become increasingly frustrated. Regarding U.S. export control policy, the former U.S. official Susan Shirk observes that "Cold War fears and protectionist instincts are clouding Americans' economic reason. Overreactions, which are read by the Chinese public and its leaders as an expression of our hostile intentions . . . could turn China from an economic rival into an all-out enemy."[34] The China expert Alanna Krolikowski offers a reciprocal rationale in observing: "While China's capabilities in space are known to U.S. observers, its intentions are not. The status quo may deprive the United States of options and tools for learning about these intentions."[35] In light of new market realities and China's growing space budget, over one hundred member companies of the Aerospace Industries Association recently signed a letter to President Obama calling for the ITAR list to be reviewed and for the administration to develop clear criteria for those sensitive technologies that are not otherwise available in the marketplace and require export control protection.[36] The administration has yet to deliver on these reforms, although it continues to move the interagency process in this direction.

Military Space

One of the greatest risks of twenty-first-century space competition at the global level is that countries might revisit the history of arms racing that characterized the run-up to World War I. Equally frightening to some analysts and officials is the threat of a World War II scenario, where one authoritarian state (today seen as China) exploits a policy of opacity and overall global policies of restraint in the deployment of space weapons to develop an asymmetric advantage in this potentially revolutionary field, which it then uses to threaten, subvert, or destroy its rivals. This is exactly the scenario outlined in the 2001 Rumsfeld Commission report, which focuses on the threat of a "space Pearl Harbor" against the United States, presumably carried out by Beijing. China's 2007 test of an ASAT weapon and its opposition to military-to-military talks on space security seem only to underline this risk.

However, while the metaphor of a "Pearl Harbor" resonates strongly among politicians of Secretary Rumsfeld's generation, changes in the world situation since then and the characteristics of space itself seem to make this a poor analogy. Given the physics of orbital space, it is not as easy for a country to launch a major attack on a constellation of assets as it might against a fixed target on Earth. Moreover, given the limitations of China's current ASAT system and the fact that many, if not most, critical U.S. space assets are located in orbits beyond the range of its ground-based missiles mean that such an attack would need to rely on a range of other technologies, some of them based in space. Such an attack would also take hours to conduct, during which time the United States would actively fight back against China's launch facilities, radars, and communications systems and take steps to avoid further interceptions (such as by moving key orbital assets). China would also be branded a rogue by the world community and likely find that its actions had rallied strong international support behind the United States. This is not to say that the United States and other countries do not face future risks to their space assets, but the scope of damage has likely been overestimated, and the ability of capabilities to be reconstituted—particularly in the presence of expanding commercial assets in such critical areas as satellite reconnaissance, communications, and, in the future, global positioning—has been underestimated.

As for China's actual motivations, while deductive arguments often posit virtually unlimited evil intentions upon Beijing's military space program, better-informed inductive analysis drawing on actual Chinese capabilities and test programs show significant shortcomings and the absence of current plans for space "dominance." As the U.S. National Defense University China expert Phillip Saunders observes, "China may ultimately settle for limited counterspace capabilities (for example, a limited capability to temporarily disrupt U.S. use of space assets)" out of its growing concern about its ability to "maintain the use of its own space assets" in a degraded space environment.[37] Thus, China is not likely to be immune to the factors that eventually resulted in space weapons restraint between Moscow and Washington, if it can be engaged in serious discussions on space security.

China has taken an active role in trying to address other transnational security problems in Asia, such as the North Korean nuclear crisis, where it has participated alongside the United States in conflict resolution with other key countries in the Six-Party Talks (North and South Korea, Japan, and Russia). While the effort has not yet halted the DPRK program's advancement, it has helped identify common U.S.-Chinese interests on this issue and has established shared principles and mechanisms for dealing with the problem. In the maritime domain, the Obama administration proposed a similar approach of international interaction and dispute resolution at the July 2010 meeting of the Association of Southeast Asian Nations in Hanoi for addressing conflicting national claims in the South China Sea.[38] However, China's rejection of this initiative a day later shows that a U.S. role may not always be welcomed, particularly on issues considered to relate to Chinese territorial sovereignty, including Taiwan. Space, however, is by UN agreement a nonterritorial zone and thus may be more conducive to cooperation.

One factor likely to assuage international military space tensions is the recent shift in U.S. space policy. In June 2010, the Obama administration released its long-awaited National Space Policy.[39] The document reversed trends under the Bush administration oriented toward asserting exclusive U.S. rights in space and pushing the envelope in the area of threatened military responses. Instead, while asserting rights to self-defense, the 2010 U.S. space policy expressed a renewed American receptivity to international cooperation and a premium on establishing new norms of noninterference and good behavior in space, in which Washington welcomed the participation of cooperative partners. As the U.S. analyst Logsdon comments, "This shift away from unilateral leadership to leadership among partners is a sea change."[40] Although critics have noted the new policy's lack of specific suggestions for space arms control, the guidelines have been widely welcomed by international observers. The changes have strengthened prospects for global adoption of "rules of the road," enhanced space situational awareness, and cooperation in identifying and sanctioning actors that violate norms of freedom of access to space by attacking, jamming, or otherwise disrupting the peaceful activities of other space actors. As President Obama explained regarding the new U.S. policy: "No longer are we racing against an adversary; in fact, one of our central goals is to promote peaceful cooperation and collaboration in space, which not only will ward off conflict, but will help to expand our capacity to operate in orbit and beyond."[41]

POSSIBLE BRIDGES: NEW APPROACHES TO COOPERATION

A key distinguishing characteristic of space activity in the early twenty-first century—compared to its first fifty years—is the increasing crowding of this new environment, which puts unprecedented pressure on space resources. The problem posed by Asia's space competition is that its timing coincides with an exacerbation of global changes in the dynamics of space activity itself with the emergence of additional states and private entities as critical players, shifting what was previously a simpler, largely U.S.-Soviet realm. The European Space Policy Institute analyst Nina-Louise Remuss explains that "The increase in actors and activities puts pressure on the available orbital and spectrum resources and the environment, and thus calls for more coordination on the basis of a space situational awareness system."[42] In this evolving space environment, countries are beginning to realize both the limits of unilateral actions and the necessity of cooperation to prevent space conflicts, which are ultimately self-defeating. As Per Magnus Wijkman argues, such conditions of interdependence give actors "strong incentives to agree on measures to keep interference at a mutually acceptable level."[43] But the challenge of how to do so—as in the field of international trade—involves complex multilateral negotiations and a willingness to compromise for the sake of a greater common good. Since this has not been the stock-in-trade of most Asian countries, new mechanisms and motivations will be needed in the three fields of space activity.

Civil Space

As discussed above, a significant challenge in moving toward a more cooperative space future in Asia is the current infrequency of regionwide space dialogue and planning. Although a few projects exist—such as Sentinel Asia, the *SMMS*, and the STAR program—these have been limited to a subset of regional countries. They are a good start but are unlikely to lead to broader cooperation and may even widen the existing regional divide between the Chinese-led APSCO and the Japanese-led APRSAF. In order to bridge this gap, the South Korean space analyst Doo Hwan Kim has proposed the phased creation of a joint APSCO-APRSAF organization that he calls the Asian Space Development Agency (ASDA).[44] Kim outlines the possible phases as (1) developing a draft convention, (2) establishing an electronic ASDA through the Internet, (3) creating a related Asian Center for Space Law, and then (4) creating a formal center and system of joint financing, as ESA has done for Europe. As Kim describes his rationale for the proposed organization: "It is desirable and necessary . . . to establish an ASDA, in order to develop our space industry, to strengthen friendly relations and to promote research cooperation among Asian countries, based on oriental ideology, ethics and creative ideas." Notably, however, Kim believes that an Asian identity for the organization is critical, as he sees space developments in the future leading to a struggle for the Asian space market between less-developed Asian space programs and the advanced programs in the United States, Russia, and ESA. As he argues: "To win this severe competition, it will be necessary for the Asian peoples to work together in union."[45] But, while possibly good for Asia in the short term, it remains questionable whether fostering such a regionally based competition with the more advanced space programs is the best approach, particularly if broader space cooperation internationally might act instead as a rising tide that lifts all boats.

A different approach is suggested by the European space analyst Remuss, who argues: "The increased pressure on the environment and the existing resources [should] be countered by fostering a dialogue between spacefaring countries and newcomers."[46] India's Moorthi agrees with this argument and emphasizes the growing importance of transnational approaches to space problems, observing: "While countries . . . still have a national perspective arising out of their own geography and history, global trends and global challenges also need to be dealt with appropriately."[47]

The space expert James Vedda outlines a supportive approach, making the case that all countries share an interest in getting civil space activity to serve "Earth-centric" interests in the coming decades rather than engaging in costly programs of competitive space exploration along the lines of the Apollo program.[48] Vedda argues that shared problems such as global warming, environmental degradation, and energy shortages provide strong incentives for a new, international perspective on space. As Vedda observes,

"Collectively, the space community, which has a tendency to be too insular, must open its intellectual aperture wider than the space visionaries of the past and seek to encompass the highest-salience global challenges of our era."[49] He emphasizes that "The knowledge-based activity that space efforts thrive on must draw from around the world."[50] Can Asian countries make this leap? If there are current obstacles, what might some of the problem areas be?

Human spaceflight is a hard case for international cooperation, because it is also one of the highest-prestige areas of spaceflight and the most costly. China has clearly invested in this area not to promote cooperation with others but to serve national goals. The same could be said for the early American and Soviet programs, but times have changed, leading to the remarkable cooperation seen in the *ISS* project since the 1990s. Today, the extremely high expected costs of next-generation exploration missions—such as settlement of the Moon or the first human spaceflight to Mars—present daunting challenges for any single nation. In this context, the logic of joint missions increases. Notably, all of the major Asian spacefaring countries—China, India, Japan, and South Korea (among other nations)—participated in meetings with NASA, Russia, and ESA to develop the Global Exploration Strategy (GES) in May 2007. The GES seeks to improve data exchanges among members undertaking planetary research and other space exploration, identify areas of both gap and overlap, and promote cooperation among the involved entities. The report specifically identifies goals of interoperability and the establishment of common standards to allow the future space-based economy to work cooperatively. All of these Asian countries have also indicated an interest in joining the NASA-led International Lunar Network, which seeks to develop cooperative mechanisms for sharing scientific data from experiments on the Moon. These and other initiatives make it increasingly likely that the next major space ventures will not be unilateral. As two experts who have studied this movement argue, "We are currently on the verge of transitioning to a new space exploration era, Space Exploration 3.0. This new phase of space exploration will involve not only states through their space agencies, but also industries, space entrepreneurs, universities and other nongovernmental organizations."[51] They conclude: "The new era of space exploration will be international, human centric, transdisciplinary and participatory."[52] Given China's significance as a future space player, even former NASA Administrator Michael D. Griffin concluded his otherwise highly pessimistic 2007 memorandum to the U.S. National Security Council about China's space capabilities with this piece of policy advice aimed at President Bush's successor: "I recommend that we engage China in such a way as to curb the tendency toward rivalry in space with the United States."[53] Senior Chinese space officials, notably, remain open to human spaceflight cooperation with the United States and even to a joint U.S.-Chinese flight,[54] perhaps similar to the symbolic Apollo-Soyuz flight that temporarily marked a warming of the U.S.-Soviet space race in 1975. One Chinese space analyst notes that, despite the recent rise of nationalism in China, even the "angry youngsters" are supportive of space cooperation with the United States, given the significance that U.S. recognition would involve.[55]

Despite these positive signs, more needs to be done. With the declining NASA interest in returning to the Moon and uncertainty about future manned exploration, the prospect for enhanced U.S.-Chinese cooperation in high-prestige missions remains murky. Whether cooperation in robotic missions can begin to ease tensions and build cooperative links between the two major space programs is a matter for higher-level attention in both capitals. While international cooperation on the *ISS* represents a model for many other countries (Russia, ESA, Japan, etc.), it has not yet included China or India. If China proceeds with its own space station, as currently planned, and seeks to develop an autonomous lunar exploration program, goals of meaningful human spaceflight cooperation may have to be shelved. Yet Russia's role in China's robotic program for the Moon, ongoing discussions through the International Lunar Network and other forums, and cost concerns for major settlement activities suggest that transnational efforts are still entirely possible.

In terms of bilateral U.S.-Chinese contacts, the best approach may be to start small. The Chinese space analyst Yi Zhou believes the Chinese Academy of Sciences would be a better partner than CNSA, because "it has a more open attitude toward international scientific cooperation and is less bureaucratic."[56] At the same time, Zhou notes the growing generational change in China's academic institutes and that the experience of many younger Chinese in either studying or working overseas gives the new generation a more positive attitude toward space cooperation than its predecessors. On the U.S. side, exempting U.S. universities from ITAR rules on most space science payloads—as has been done with India—could make a major difference in jumpstarting cooperative projects.

Commercial Space

While many leading powers have entered space for military reasons or as spinoffs from missile programs, most countries become involved in space activities for primarily commercial reasons: specifically, to help develop their economies and bring services to their populations. With the exception of China, all of the Asian space programs currently involved in space activity began with a major emphasis on these activities, helping to drive what has now become a $189-billion-a-year industry worldwide.[57] In sharp contrast to the situation during the cold war, much of this commerce takes place across national borders and involves companies whose ownership may be in one country, whose facilities may be in another, whose technologies may come from yet other countries, and whose services are marketed to still others. Thus, the globalization of the space industry is an emerging fact, even if the U.S. Congress is fighting this process alone.

The announcement in June 2010 by then-U.S. National Security Council Advisor James L. Jones that the Obama administration was seeking to create a single licensing agency for exports that would be independent of both the U.S. Departments of State and Commerce, however, suggests that changes may be forthcoming to enable greater participation by U.S. companies in the global space market, including forging new ties with Asian economies.[58] James Lewis argues that "The goal should be to accelerate innovation in the US rather than continue efforts to slow innovation in China."[59] This latter strategy has clearly not worked, since the United States no longer is the sole producer of advanced satellites for the international market.

While some analysts continue to argue that space commerce will eventually lead to military conflict, the way it did in the late nineteenth-century grab for colonies, or that China represents another Wilhelmite Germany disrupting the peace established by the Concert of Vienna,[60] many other specialists on Asia cast doubt on these predictions, given important changes in the international system. As retired U.S. Admiral Dennis Blair argues, "Although the rise of China is often compared to the rise of

Germany in the late nineteenth century, the differences are more compelling than the similarities."[61] The markers of state power have shifted since that period away from territorial acquisition via force to commercial assets and national prosperity, in a world where warfare is no longer considered acceptable or even desirable among the major powers, given the tremendous destructive capacity of modern weapons. Also, the far greater economic integration of the world's economies and financial institutions has changed the incentive structure facing political elites, adding new constraints. James Mulvenon observes cogently: "With the abandonment of Communist ideology, the basis of CCP legitimacy has shifted to the continued deliverance of economic prosperity to the population."[62] Indeed, throughout Asia, Blair argues, leaders are becoming increasingly cognizant that "military tension and wars are bad for international business."[63]

Ironically, recent tests of military systems in space as well as commercial accidents have strengthened norms of restraint, at least outside Asia. Specifically, the debris-generating effects of the China's 2007 ASAT test and the 2009 Iridium-Cosmos collision have resulted in new international pressure among commercial users for international cooperation in debris prevention. In fact, a coalition of commercial users of geostationary orbit has created the first private space sitvational awareness system.[64] This group includes the three leading users of this top orbital belt—Intelsat, SES, and Inmarsat—that together have established the Space Data Association, based in the United Kingdom, to track geostationary objects and share data among users in order to prevent collisions or harmful interference. Their meetings in 2010 to craft joint policies to deal with the threat posed by the drifting and unresponsive *Galaxy 15* satellite marked a significant first step toward commercial "policing" of space in advance of governmental collaboration. The U.S. government has also instructed the U.S. Air Force to develop and share additional "conjunction analysis" data for non-U.S. spacecraft from its Space Surveillance Network database in order to prevent future spacecraft collisions. These new forms of collective action suggest that international management of space may yet be possible. Formation of a broader council of national space agencies could continue this trend and begin to create legal mechanisms for enforcing best practices and collective responses to emerging space problems.

A similar approach to expanding joint development might be to focus on international space infrastructure and shared utilities. The widespread use of the U.S. GPS network is currently spurring copycat systems in Europe, Russia, India, and China, as well as smaller augmentation systems in Japan and South Korea, for fear of a future loss of signals should the U.S. military decide to shut off this critical information to commercial users. Although the United States has provided this service for free to international users since it began providing unencrypted signals for civilian use in the late 1980s and has no specific plans to cease this service, U.S. declarations of an intention to integrate this system with other international efforts and to guarantee its accessibility (short of hostile action) could assuage concerns. It could also build collective incentives to ensure the system's continued operation, such as via national pledges of noninterference from major spacefaring nations and more formal, international legal mechanisms. Joined with a council of space operators, such an approach might begin to focus on collective development of space as an infrastructure project, similar to the creation of the U.S. interstate highway system in the 1950s or the establishment of international rules for air travel. The space analyst J. M. Snead calls upon the United States to lead a process focusing "aerospace industries on building and operating an integrated spacefaring logistics infrastructure and, then, using the newly acquired mastery of human space operations to enable the emergence of a new generation of commercial space enterprises."[65] While Snead outlines a proposed $200 billion U.S. effort as the key starting point, space commercial dynamics suggest that an *international* effort would not only be more feasible but also build common interests in mutual protection of this infrastructure and to stimulate further transnational commercial cooperation. Eventually, in Snead's terminology, such an effort could "progress to resource extraction, local industrialization, settlement, and interconnected economic and social development."[66] Through these means, commercial space conflict need not arise but instead might become a force facilitating a growing international infrastructure for the use and development of space.

Military Space and Space Security Proposals

A number of leading Asian space actors are on record supporting a strengthening of the regime for space security. All of Asia's spacefaring countries have voted consistently for UN resolutions on preventing an arms race in outer space and on increasing transparency. But these UN statements have not resulted in much progress for specific measures. Foreign Minister Yang Jiechi stated in a speech in Geneva in August 2009, "Credible and effective multilateral measures must be taken to forestall the weaponization [of] and arms race in outer space."[67] Yang implicitly criticized the United States by saying, "Countries should neither develop missile defense systems that undermine strategic stability nor deploy weapons in outer space."[68] But although Beijing and Moscow still hold out hope for salvaging their proposed PPWT, international support has cooled substantially, and the agreement has no serious chances of moving forward. This leaves the European "Code of Conduct" as the only viable space security measure under active consideration, but this proposal has not yet been introduced formally into an international body. This is a rather sorry state of affairs, given the seriousness of the emerging risks to space security posed by loopholes in existing treaties.

Meanwhile, the Conference on Disarmament (CD) remains in a stalemate, with no progress on space talks, much less actual negotiations. In this environment, rhetoric dominates, and real progress is delayed. Today, Washington's tone is more moderate than it was under the Bush administration, but this has not resulted in actual diplomatic progress toward any new agreements. Instead, military testing in space continues, fostering tit-for-tat responses from a slowly growing cascade of countries. As Johnson-Freese argues: "A free-for-all is where we are headed if a course correction is not made soon."[69] However, it takes two to reach a compromise, and China has avoided direct military-to-military space talks. But is China perhaps reluctant to speak because of the U.S. lead in space technology and its fear of the comparative gap?

Deborah Larson and Alexei Shevchenko have argued recently that one of the reasons for China's unwillingness to engage the United States in serious military-to-military discussions may be its concerns about status and past feelings of inferiority.[70] Countries that believe their strategic interests lie in obfuscation and secrecy, they argue, may not be concealing secret military programs of tremendous power but instead weaknesses that they do not wish to reveal to the outside world. Eligar Sadeh identifies similar tendencies in China that affect its space-related behavior, arguing, "The self-image of China is one that is inferior in power and force projection to the United States. This engenders for the Chinese a sense of strategic vulnerability."[71] Such concerns clearly motivated Soviet behavior during the test-ban negotiations in the 1950s, when the Soviet arsenal lagged considerably behind that of the United States and caused Moscow to reject on-site inspections of Soviet nuclear facilities.

Indeed, similar arguments might be made of a number of Asian powers in space, certainly including at least India and the two Koreas, if not Japan as well. Sadeh argues that a U.S. tendency to frame space challenges in largely a military context may exacerbate these concerns. As he argues: "China's military thinking is directly affected by U.S. policy and programs directed at space power for space dominance and control. This also leads to insecurity among Chinese leaders."[72] These points suggest that a special effort by the United States to include Asian countries in future space security discussions and to give them a seat at the table in the formulation of a new strengthened regime for space may bear fruit—specifically, by giving them the status they seek as equals. Another approach, advocated by Sadeh, is to reorient the focus of U.S.-Chinese space discussions, at least in the near term, away from military issues and instead toward civil space cooperation as a means of building confidence on the Chinese side concerning the benefits of collaboration with the United States.[73]

But getting to square one in space security talks may present the biggest challenge. Thus, how might Asia get started on this process without feeling that it is being taken advantage of by the more advanced space powers? One point that emerges from the U.S.-Soviet space relationship is the early establishment of a norm of noninterference with each other's spacecraft. Thanks to the arms control process in the early 1970s, the two sides were able to codify this norm into a formal agreement not to interfere at least with each other's reconnaissance and early-warning satellites. Given widespread support among Asian capitals for the concept of political noninterference and for space transparency, as well as the recent inclusion of noninterference with spacecraft in the 2010 U.S. National Space Policy, this issue could be a very good starting point. Such an initiative could be made at the CD in Geneva, at the UN First Committee in New York, at the UN Committee on the Peaceful Uses of Space in Vienna, perhaps at a special joint meeting of APRSAF and APSCO (at which the United States, Russia, and ESA might participate), or simply on a bilateral level, starting with the United States and China. It would be difficult for any country to refuse this initiative—as it would clearly expose any hostile intentions—and it would begin to identify common ground among spacefaring states in developing stronger norms for space security. Such moves might then lead to broader discussions—as called for the U.S. National Space Policy—"to identify, locate, and attribute sources of radio frequency interference" and to do so "in cooperation with civil, commercial, and foreign partners."[74] This could begin to build a system of joint management of, and responsibility for, space.

Another serious problem related to confidence building and debris mitigation is the continued testing of kinetic and nonkinetic weapons in space and the absence of related controls. The use of ASATs by both China and the United States in the past few years and increasing fears of laser weapons or other directed energy systems being used against space objects make this an important area for progress, if future space security is to be assured. There is widespread agreement that tests against space objects (particularly those above approximately 180 kilometers) create dangerous debris. Laser tests are usually less obviously damaging but can cause fuel tanks to explode and other equipment to fracture, leading to debris generation. While the normal preference of national militaries is to promote freedom for such research, there is a serious cost of continuing to damage the space environment. To date, however, no country has challenged others to stop this harmful process by declaring a unilateral moratorium and inviting others to join it—and perhaps creating an international monitoring group. Such an initiative, by especially China, India, Russia, or the United States, could go a long way toward slowing the current action-reaction dynamics in space weapons development and begin to foster common interests in collective "policing" of space by responsible space users. Over time, such efforts could lead to new forms of engagement among national militaries to ensure common rights to derive information from space, while creating strong norms and future enforcement mechanisms to prevent new actors from carrying out dangerous activities.

One problem linked to space testing and other military activities is the failure of a number of countries to comply fully with existing space law. A case in point is the 1975 UN Registration Convention, which calls on states to provide information on their space-launch activities, including the basic orbital parameters and the general function of the spacecraft. For many years, members of the regime have tolerated partial declarations, even though these data are becoming increasingly important for space traffic management and control. Moreover, there is currently no requirement for countries to inform any international body when they move their satellites, thus creating inevitable hazards and raising the costs of monitoring space traffic. A concerted effort to address compliance issues with the Registration Convention both regionally and globally, while also building in new provisions for strengthened data sharing (similar in certain ways to the International Atomic Energy Agency's Additional Protocol), would help reduce problems in space traffic management and increase transparency among states harboring mistrust of their neighbors, such as in Asia. Over time, such a system would also reduce incentives for cheating and for launching experimental military spacecraft that might be exposed. Again, such a system would help reduce prospects for harmful space "breakout" by any country, thus increasing mutual space security.

A major concern that has kept the United States from cooperating more with China in space is past Chinese cooperation with third countries in missile development—including Pakistan, Iran, and North Korea. While China has pledged to abide by the terms of the Missile Technology Control Regime (MTCR) and not transfer technologies that would assist these countries in developing missiles with ranges longer than three hundred kilometers and payloads heavier than five hundred kilograms, the U.S. ability to determine Beijing's current compliance has been hampered by lack of transparency on the Chinese side and an unwillingness to discuss these matters. A concerted attempt by both sides to bring China into the MTCR might be one means of overcoming this mistrust while also improving its compliance. On the U.S. side, such an initiative would mean sharing sensitive intelligence data with China about foreign missile proliferation concerns. On the Chinese side, greater transparency on the activities of domestic companies and a willingness to allow Chinese export control officials to meet with their American (or other MTCR member) counterparts would be required. With China working cooperatively with other MTCR members to enforce missile nonproliferation, an important side benefit is that the United States could be more confident of the security of its technology in working with China in space cooperation. This confidence-building measure and experience in joint missile control efforts would increase mutual knowledge of each other's governmental systems and give Beijing a sense that it is on the same side as Washington in seeking to pursue common nonproliferation goals.

In order to help defuse regionwide military tensions, the Japanese space scholar Kazuto Suzuki has proposed a regional, cooperative space monitoring program—which he calls the Transparency Information Center—as one such possible mechanism.[75] With Japanese help, the proposed organization would offer a Regional Confidence-Building Satellite to "establish an early-warning center to provide imagery intelligence data and to increase transparency regarding troop mobilizations and proliferation monitoring."[76]

But other analysts see emerging, nontraditional security problems as best dealt with through interaction with relevant outside players. As Caballero-Anthony argues, "when and where external help and expertise are required, countries outside the region should be welcome participants."[77] Many analysts believe that the most expedient and effective outcome for space is to rise above the regional level and create an expanded international pact to fill gaps in the 1967 Outer Space Treaty. The German space analyst and official Detlev Wolter argues that the best option for mankind in space is "The conclusion of a multilateral agreement and the creation of an international organization for . . . common security in outer space."[78] His treaty proposal includes such provisions as a multilateral space monitoring system, the destruction of all ASAT systems, and a comprehensive ban "on military uses of a destructive nature" in space.[79] Yet despite its good intentions and its seeming support of key space themes voiced at various times by Japan, India, and China, the treaty has thus far not attracted their official backing and has not moved forward in any arms control negotiating forum.

In the end, as Emily Goldman and Andrew Ross caution about trends toward space's weaponization: "More may be lost than gained by extending the terrestrial security dilemma into space."[80] Thus, while the specific starting point desired by individual Asian countries may vary, the overriding point is the need to get started in finding a solution.

CONCLUSION

Moving off of the Earth poses a challenge to nation-states. National governments have traditionally been the entities that have funded space activity, through programs that have often been motivated by competitive aims.

These trends certainly characterized the cold war and the space race of the 1960s. The Indian analyst Ajey Lele observes that, in the twenty-first century, "Asia is rapidly emerging as a 'sphere of influence' in the space arena" and that nations in the region understand that "the inherent dual-use nature of space technology increases options for states to use it for militaristic purposes."[81] The trends toward competition are undeniable. For national military leaders, the easiest course of action may be to drive budget increases by resorting to hostile nationalism and worst-case thinking, even at the cost of exacerbating space security and relations with other Asian countries more generally.

At the same time, this very type of space competition is becoming increasingly dysfunctional to the solution of shared problems in space caused by the growing number of actors and the shrinking of finite resources in near-Earth space. Thus, cooperative efforts have become more imperative, even as Asia's space race has propelled a variety of new actors, with competitive aims, into this new environment. A question that Asian capitals face is whether the slow learning curve followed by the United States and the Soviet Union toward eventual competitive restraint can be adopted and made substantive enough to solve their emerging problems. The European space analyst Remuss speaks of the need for rapid learning "in order to avoid newcomers making the mistakes of the earlier users."[82] But space is unlikely to solve problems in isolation from the state of relations in other areas. Thus, the challenge comes back to one of breaking emerging patterns toward confrontation and instead creating dialogues in which issues like space can be broached.

The U.S. statesman Henry Kissinger commented on the nature of Western-Asian ties in an essay published on the eve of the Hu Jintao–Obama summit in January 2011, in which he contrasted the U.S. tendency to look for quick solutions of specific problems with the Chinese preference for broader approaches and a focus on long-term stability. The crux of the problem, he argued, is the absence between the two capitals of "an overarching concept for their interaction."[83] In the end, Kissinger concluded: "A concept of a Pacific community could become an organizing principle of the 21st century to avoid the formation of blocs."[84] Including space within such a framework is imperative.

Certain trends suggest that the unique characteristics of space make it a good candidate for promoting a sense of community among nations, at least once this process of talking has been initiated. As the Indian expert Moorthi argues, "space has provided a perspective to humanity to view the Earth as a total system."[85] Moorthi argues that such a "human security" approach to space's future development could allow countries to change existing patterns of Asian behavior and perhaps to bridge the gap with outside powers. But such a shift in mindset is far from inevitable, despite its logic.

Robert Jervis cautions proponents of state learning that "It is . . . not true that an increase in knowledge necessarily increases the actor's ability to cope with his environment."[86] Yet in conditions where *self-interested* incentives promote international cooperation, as today in space, such knowledge may be more likely to facilitate collective outcomes. There is a growing understanding among Asia's spacefaring nations of the existence of *common* threats to space security (such as orbital debris) and the bene-fits of cooperation in such fields as disaster prevention and environmental monitoring. So far, encouraging steps have been taken to begin dealing with these problems, and new forms of cooperation have emerged. As the political scientist Robert Keohane argues: "To pursue self-interest does not require maximizing freedom of action. On the contrary, intelligent and farsighted leaders understand that attainment of their objectives may de-pend on their commitment to institutions that make cooperation possi-ble."[87] Specific action is now required, particularly regarding space. The alternative may be the risk of losing safe access to space altogether, which is not in anyone's interests.

The purpose of this book has been to highlight harmful trends in Asia's space competition and to explain and analyze the motivations of Asia's emerging space programs. Although it has voiced concern regarding the potential for future conflict, its conclusion remains one of guarded opti-mism. Current trends toward conflict *are* avoidable, if leaders take reason-able measures to address problems concerning space activity before they fester, sow mistrust, and become tainted by nationalist sentiments that box in decision makers. Fortunately, countervailing pressures encouraging cooperation from the space environment, cost factors, and broader pro-cesses of economic globalization may yet temper drives for unilateral or military-based solutions. If Asian space powers can begin to recognize the risks of space conflict and take more purposeful action to prevent it, they still have time to head off potential problems.

Given the weight of Asia's new powers in global space affairs, the prospects for successful regional space management will go a long way toward determining the future of international relations in space more generally. Thus, getting this right is important. In this and other fields of nontraditional security, Singapore's Caballero-Anthony reinforces this point by arguing: "Asia's new regionalism has reached a critical point."[88] While various forms of Asian nationalism proved to be disruptive forces in the past century, the skill of Asian leaders in crafting new forms of multilateralism may decide the viability of Asia's future space prosperity and security in the twenty-first century and beyond.

NOTES

INTRODUCTION: ASIA'S NEW PRESENCE IN SPACE

1. Bill Emmot, *How the Power Struggle Between China, India, and Japan Will Shape Our Next Decade* (New York: Harcourt, 2008), 13.

2. See the European Space Agency Web site, http://www.esa.int/SPECIALS/About_ESA/SEMW16ARR1F_0.html.

3. On the recent impact of national budget problems on ESA's space budget, see Andy Pasztor, "European Space Goals Come Back to Earth," *Wall Street Journal* (August 9, 2010): A14.

4. See, for example, Laurence Nardon, "Europe Chooses Peace in Space" (op-ed), *Space News* (February 9, 2009): 19. See also Gerard Brachet, "Collective Security in Space: A Key Factor for Sustainable Long-Term Use of Space"; Detlev Wolter, "Common Security in Outer Space and International Law: A European Perspective"; and Xavier Pasco, "Toward a Future European Space Surveillance System: Developing a Collaborative Model for the World," all in John M. Logsdon, James Clay Moltz, and Emma S. Hinds, eds., *Collective Security in Space: European Perspectives* (Washington, D.C.: Space Policy Institute, George Washington University, January 2007).

5. See the obstacles facing Asian space cooperation as outlined in Masashi Nishihara, "The Changing Asian Security Environment"; Dipankar Banerjee, "Indian Perspectives on Regional Space Security"; Zhong Jing, "China and Space Security"; and K. K. Nair, "Space Security: Reassessing the Situation and Exploring the Options," in Logsdon and Moltz, eds., *Collective Security in Space: Asian Perspectives*.

6. Setsuko Aoki, "Japanese Perspectives on Space Security," in Logsdon and Moltz, eds., *Collective Security in Space: Asian Perspectives*, 62.

7. Gordon G. Chang, "The Space Arms Race Begins," *Forbes* (November 5, 2009).

8. Ibid.

9. Saibal Dasgupta, "No Plans for Arms Race in Outer Space," *Times of India* (November 5, 2009), http://timesofindia.indiatimes.com/articleshow/msid-5200398,prtpage-1.cms. See also Peter J. Brown, "Space Is Suddenly on the Agenda," *Asia Times* (November 12, 2009), http://www.atimes.com/atimes/China/KK12Ad01.html.

10. See, for example, Susan L. Shirk, *China: Fragile Superpower* (New York: Oxford University Press, 2008).

11. Michael D. Griffin, NASA Administrator, memorandum on "Chinese Lunar Capabilities" (November 13, 2007), 1.

12. Ibid.

13. Ibid.

14. Ibid.

15. See, for example, Frank H. Winter, "The Genesis of the Rocket in China and Its Spread to the East and West," in A. Ingemar Skoog, ed., *History of Rocketry and Astronautics*, vol. 10 of the *Proceedings of the Twelfth, Thirteenth, and Fourteenth History Symposia of the International Academy of Astronautics* (San Diego, Calif.: American Astronautical Society, 1990). See also Fang-Toh Sun, "Early Rocket Weapons in China," in *History of Rocketry and Astronautics*, vol. 14 of the *Proceedings of the Eighteenth and Nineteenth History Symposia of the International Academic of Astronautics* (San Diego, Calif.: American Astronautical Society, 1993).

16. This book uses the term "civil" space to denote nonmilitary, government-run space activities, such as space science, remote sensing, and exploration. This concept is distinct from "civilian," which denotes nongovernmental space activities, and "commercial," which designates activities conducted for profit, most often by private companies.

1. ASIAN SPACE DEVELOPMENTS: MOTIVATIONS AND TRENDS

1. William E. Burrows, *This New Ocean: The Story of the First Space Age* (New York: Random House, 1999).

2. On this history, see James Clay Moltz, *The Politics of Space Security: Strategic Restraint and the Pursuit of National Interests* (Stanford, Calif.: Stanford University Press, 2008).

3. For technical details on orbital debris and EMP effects, see David Wright, Laura Grego, and Lisbeth Gronlund, *The Physics of Space Security: A Reference Manual* (Cambridge, Mass.: American Academy of Arts and Sciences, 2005).

4. On this history, see Paul B. Stares, *Space and National Security* (Washington, D.C.: Brookings Institution, 1987).

5. On this phenomenon in the U.S. and Soviet decisions to deploy the hydrogen bomb in the 1950s, see Herbert F. York, *The Advisors: Oppenheimer, Teller, and the Superbomb* (Stanford, Calif.: Stanford University Press, 1976), 102.

6. Colonel (U.S. Army, ret.) Larry M. Wortzel, "The Chinese People's Liberation Army and Space Warfare," *Astropolitics* 6, no. 2 (May–August 2008): 128.

7. Ibid.

8. See, for example, "U.S. House Members Blast China," *Associated Press* (May 2, 2007).

9. John J. Tkacik, "China Space Program Shoots for the Moon," *Washington Times* (January 8, 2010), http://www.washingtontimes.com/news/2010/jan/08/china-eyes-high-ground.

10. Roger Handberg and Zhen Li, *Chinese Space Policy: A Study in Domestic and International Politics* (New York: Routledge, 2007), 36.

11. Robert Jervis, *Perception and Misperception in International Politics* (Princeton, N.J.: Princeton University Press, 1976), 24.

12. In fact, because of its timing, the shuttle decision (and its misperception by the Soviets) may have done more to impose costs on the Soviet military than the 1983 Strategic Defense Initiative.

13. James A. Lewis, "China as a Military Space Competitor," in John M. Logsdon and Audrey M. Schaffer, eds., *Perspectives on Space Security* (Washington, D.C.: Space Policy Institute, George Washington University, December 2005), 113.

14. Jervis, *Perception and Misperception in International Politics*, 24.

15. Ibid., 212.

16. *Comprehensive Report of the Special Advisor to the DCI on Iraq's WMD* (Washington, D.C.: U.S. Government Printing Office, September 30, 2004), http://www.gpoaccess.gov/duelfer/index.html.

17. Ibid.

18. John Arquilla, *Worst Enemy: The Reluctant Transformation of the American Military* (Chicago: Ivan R. Dee, Publisher, 2008), 99.

19. Gregory Kulacki and Jeffrey G. Lewis, "Report: Understanding China's Antisatellite Test," *The Nonproliferation Review* 15, no. 2 (July 2008): 335–336.

20. Ibid., 337.

21. Joan Johnson-Freese, *Space as a Strategic Asset* (New York: Columbia University Press, 2007), 11.

22. Thorstein Veblen, *Imperial Germany and the Industrial Revolution* (1915; repr. Ann Arbor: University of Michigan Press, 1966), 24.

23. Ibid., 41, n. 1.

24. John Arquilla, "Patterns of Commercial Diffusion," in Emily O. Goldman and Leslie C. Eliason, eds., *The Diffusion of Military Technology and Ideas* (Stanford, Calif.: Stanford University Press, 2003), 358.

25. Ibid.

26. Maj. Gen. James B. Armor Jr., "Viewpoint: It Is Time to Create a United States Air Force Space Corps," *Astropolitics* 5, no. 3 (September–December 2007): 275.

27. See Alexander Gerschenkron, *Economic Backwardness in Historical Perspective: A Book of Essays* (Cambridge, Mass.: Harvard University Press, 1962).

28. Ibid., 29–30.

29. Ibid., 29.

30. On these points, Handberg and Li stress the *similarities* between China's program and those of the Soviet Union and the United States. See Handberg and Li, *Chinese*

Space Policy, 13. My argument stresses the *differences*, based on the significant gap in China's space development and the fact that its real space effort did not begin until the late 1970s. I also point out that China benefited significantly from outside technology in a way that neither the U.S. or Soviet programs did.

31. On this argument, see Johnson-Freese, *Space as a Strategic Asset.*

32. Handberg and Li divide spacefaring nations instead into three categories: first, second, and third wave. See *Chinese Space Policy*, 39–47. They start with the first wave of programs with roots in the 1950s, counting the United States and the Soviet Union. They place China on the trailing edge of this wave, while placing Japan in the second wave, programs with roots in the 1960s and early 1970s. They then place India, South Korea, and other Asian spacefaring countries in the third wave (late 1970s and beyond). My analysis instead uses only two categories, because of its emphasis not on the year of initiation but on those countries (the United States and Soviet Union) that developed the first space technologies and proved all of the basic concepts. The second-generation states benefited from this technology and information. It thus places all of Asia's space-faring countries in the second-generation category, since all Asian countries drew significantly on existing technology and know-how developed elsewhere. This does not denigrate their technological accomplishments but recognizes that knowing what was possible mattered in giving them specific advantages as latecomers.

33. Raju G. C. Thomas, "India's Nuclear and Space Programs: Defense or Development?" *World Politics* 38, no. 2 (January 1986).

34. Ibid., 317.

35. Kartik Bommakanti, "Indian Space Launch Vehicles and ICBM," report for the Space Security Project, Web site of the Center for Defense Information (February 29, 2008), http://www.cdi.org/pdfs/IndiaICBM.pdf.

36. James A. Lewis, "Space and Power in 2007," *Space News* (December 17, 2007): 25.

37. Ibid.

38. Michael Mastanduno, "Hegemonic Order, September 11, and the Consequences of the Bush Revolution," in G. John Ikenberry and Chung-in Moon, eds., *The United States and Northeast Asia: Debates, Issues, and New Order* (Lanham, Md.: Rowman and Littlefield, 2008), 263.

39. Ibid.

40. Ibid., 264.

41. Ashley J. Tellis, "China's Space Weapons," *Wall Street Journal* (July 23, 2007): A15.

42. Ashley J. Tellis, "Preserving Hegemony: The Strategic Tasks Facing the United States," in Ashley J. Tellis, Mercy Kuo, and Andrew Marble, eds., *Strategic Asia 2008–09, Challenges and Choices* (Seattle, Wash.: National Bureau of Asian Research, 2008), 11.

43. Tellis, "China's Space Weapons."

44. Ibid.

45. Avery Goldstein, "Power Transitions, Institutions, and China's Rise in East Asia," in G. John Ikenberry and Chung-in Moon, eds., *The United States and Northeast Asia: Debates, Issues, and New Order* (Lanham, Md.: Rowman and Littlefield, 2008), 63.

46. Bruce Cumings, "On the History and Practice of Unilateralism in East Asia: Or, the More Things Change, the More They Remain the Same," *Pacific Focus* 33, no. 2 (August 2008).

47. Ibid.

48. Squadron Leader (Indian Air Force) K. K. Nair, *Space: The Frontiers of Modern Defence* (New Delhi: Knowledge World, 2006), 117.

49. Ibid.

50. Major General (Indian Army, ret.) Dipankar Banerjee, "Indian Perspectives on Space Security," in John M. Logsdon and James Clay Moltz, eds., *Collective Security in Space: Asian Perspectives* (Washington, D.C.: Space Policy Institute, George Washington University, January 2008).

51. Ibid., 130.

52. Ibid., 120.

53. "India Readying Weapon to Destroy Enemy Satellites: Saraswat," Indiaexpress. com (January 3, 2009), http://www.indianexpress.com/news/india-readying-weapon -to-destroy-enemy-satellites-saraswat/562776/.

54. Kenneth B. Pyle, *Japan Rising: The Resurgence of Japanese Power and Purpose* (New York: PublicAffairs, 2007).

55. Ibid., 373.

56. Ibid.

57. See Gilbert Rozman, *Northeast Asia's Stunted Regionalism: Bilateral Distrust in the Shadow of Globalization* (New York: Cambridge University Press, 2004).

58. Gilbert Rozman, "South Korea and Sino-Japanese Rivalry: A Middle Power's Options Within the East Asian Core Triangle," *Pacific Review* 20, no. 2 (June 2007): 214.

59. Michael D. Swaine, "Managing China as a Strategic Challenge," in Ashley J. Tellis, Mercy Kuo, and Andrew Marble, eds., *Strategic Asia 2008–09: Challenges and Choices* (Seattle, Wash.: National Bureau of Asian Research, 2008), 96.

60. Ibid., 100.

61. See Robert O. Keohane, *After Hegemony: Cooperation and Discord in the World Political Economy* (Princeton, N.J.: Princeton University Press, 1984).

62. Thomas L. Friedman, *The World Is Flat: A Brief History of the Twenty-first Century* (New York: Farrar, Strauss and Giroux, 2006).

63. Keohane, *After Hegemony*, 258.

64. Vinod K. Aggarwal and Min Gyo Koo, "An Institutional Path: Community Building in Northeast Asia," in G. John Ikenberry and Chung-in Moon, eds., *The United States and Northeast Asia: Debates, Issues, and New Order* (Lanham, Md.: Rowman and Littlefield, 2008).

65. Ibid., 295.

66. Ibid., 299.

67. Jonathan Pollack, "US Strategies in Northeast Asia: A Revisionist Hegemon," in Byung-Kook Kim and Anthony Jones, eds., *Power and Security in Northeast Asia: Shifting Strategies* (Boulder, Colo.: Lynne Rienner, 2007), 92.

68. Muthiah Alagappa, "Introduction: Predictability and Stability Despite Challenges," in Muthiah Alagappa, ed., *Asian Security Order: Instrumental and Normative Features* (Stanford, Calif.: Stanford University Press, 2003), 18.

69. Ibid.

70. Ibid.

71. Joan Johnson-Freese, *Heavenly Ambitions: America's Quest to Dominate Space* (Philadelphia: University of Pennsylvania Press, 2009), 133.

72. Ibid., 144.

73. Ibid., 145.

74. David C. Kang, *China Rising: Peace, Power, and Order in East Asia* (New York: Columbia University Press, 2007), 4, 5.

75. Chung-in Moon and Seung-Won Suh, "Identity Politics, Nationalism, and the Future of Northeast Asian Order," in G. John Ikenberry and Chung-in Moon, eds., *The United States and Northeast Asia: Debates, Issues, and New Order* (Lanham, Md.: Rowman and Littlefield, 2008).

76. Ibid., 194.

77. Ibid.

78. Minxin Pei, "China's Hedged Acquiescence: Coping with US Hegemony," in Byung-Kook Kim and Anthony Jones, eds., *Power and Security in Northeast Asia: Shifting Strategies* (Boulder, Colo.: Lynne Rienner, 2007), 103.

79. As Veblen observed, "the carrying-over of such a state of the industrial arts from one community to another need not involve the carrying-over of this its spiritual complement. Such is particularly the case where the borrowing takes place across a marked cultural frontier, in which case it follows necessarily that the alien scheme of conventions will not be taken over intact in taking over an alien technological system, whether in whole or in part" (Veblen, *Imperial Germany and the Industrial Revolution*, 37).

80. Susan L. Shirk, *China: Fragile Superpower* (New York: Oxford University Press, 2008), 255.

81. Ibid., 267.

82. Francis Fukuyama, "The End of History?" *The National Interest* (Summer 1989). Reprinted in Richard K. Betts, ed., *Conflict After the Cold War: Arguments on Causes of War and Peace* (New York: Longman, 2008), 13, 17.

83. Ibid., 11–12.

84. Paul Midford, "Challenging the Democratic Peace? Historical Memory and the Security Relationship Between Japan and South Korea," *Pacific Focus* 23, no. 2 (August 2008): 201.

85. Stephan Haggard, "US Influence in a Changing Asia," in Byung-Kook Kim and Anthony Jones, eds., *Power and Security in Northeast Asia: Shifting Strategies* (Boulder, Colo.: Lynne Rienner, 2007), 46.

86. Ibid., 48.

2. THE JAPANESE SPACE PROGRAM: MOVING TOWARD "NORMALCY"

1. Saadia M. Pekkanen, *Picking Winners? From Technology Catch-up to the Space Race in Japan* (Stanford, Calif.: Stanford University Press, 2003), 181.

2. Kenneth B. Pyle, *Japan Rising: The Resurgence of Japanese Power and Purpose* (New York: PublicAffairs, 2007), 373.

3. Saadia M. Pekkanen and Paul Kallender-Umezu, *In Defense of Japan: From the Market to the Military in Space Policy* (Stanford, Calif.: Stanford University Press, 2010), 20.

4. Hirotaka Watanabe, "Japan's National Space Strategy: A New Diplomatic Challenge?," paper presented at the National Space Strategy Workshop, Space Policy Institute, George Washington University, Washington, D.C. (February 4–5, 2010), 3.

5. Frank H. Winter, *Prelude to the Space Age, The Rocket Societies: 1924–1940* (Washington, D.C.: Smithsonian Institution Press, 1983), 111.

6. On this history, see Brian Harvey, *The Japanese and Indian Space Programmes: Two Roads Into Space* (London: Springer, 2000), 3–5.

7. Bernard Millot, *Divine Thunder: The Life and Death of the Kamikazes* (New York: McCall Publishing Company, 1970), 124.

8. Ibid., 127.

9. Ibid., 125.

10. M. G. Sheftall, *Blossoms in the Wind: Human Legacies of the Kamikaze* (New York: Penguin, 2005), 141.

11. Harvey, *The Japanese and Indian Space Programmes*, 3. The so-called *Shusui* aircraft was designed to fly to 12,000 feet under rocket power, attack its target aircraft, and return to base as an unpowered glider.

12. Ibid., 5.

13. On the process of Japan's shift from military occupation to renewed independence, see chapter 15, "The Years of Occupation," in James L. McClain, *Japan: A Modern History* (New York: W. W. Norton and Company, 2002), 523–561.

14. Harvey, *The Japanese and Indian Space Programmes*, 5–6.

15. On this point, see Hideo Itokawa, "Space Exploration in Japan, 1960–62," in *Proceedings of the International Congress "The Man and Technology in the Nuclear and Space Age,"* Milan, Italy, April 18–21, 1962 (Rome: Associazione Internatiozionale Uomo Nello Spazio, 1963), 27.

16. Japan Aerospace Exploration Agency (JAXA), "Timeline of ISAS," http://www.isas.jaxa.jp/e/about/history/index.shtml; also Itokawa, "Space Exploration in Japan, 1960–62," 6.

17. Harvey, *The Japanese and Indian Space Programmes*, 8; JAXA, "History of Japanese Space Research."

18. Harvey, *The Japanese and Indian Space Programmes*, 8. These launches were conducted very inexpensively. The nineteen K-6 flights cost $215,000 altogether, and the subsequent nine K-8 flights in 1960 and 1961 cost $190,000. See Itokawa, "Space Exploration in Japan, 1960–62," 32.

19. JAXA, "History of Japanese Space Research."

20. Watanabe, "Japan's National Space Strategy," 2.

21. See *Proceedings of the Eighth International Symposium on Space Technology and Science* (Tokyo: Agne Publishing, 1969).

22. H. Itokawa, "Japanese Sounding Rocket Program, 1963–1965," in *Proceedings of the Sixth International Symposium on Space Technology and Science* (Tokyo: Agne Publishing, 1966), 17.

23. Itokawa, "Space Exploration in Japan, 1960–62," 28.

24. Steven Berner, "Japan's Space Program: A Fork in the Road?" RAND Technical Report (Santa Monica, Calif., 2005).

25. Harvey, *The Japanese and Indian Space Programmes*, 13. Harvey (17) notes that, following his resignation from ISAS, Itokawa served as an advisor to the emerging Indian space program until 1971.

26. Masao Yamanouchi and Yukihiko Takenaka, "Space Development Activities in Japan," in *Proceedings of the Fourteenth International Symposium on Space Technology and Science* (Tokyo: Agne Publishing, 1984), 65; Watanabe, "Japan's National Space Strategy," 3.

27. Pekkanen, *Picking Winners*, 173.

28. Akiyoshi Matsura and Yasuhiro Kuroda, "Progress Report on NASDA Projects," in *Proceedings of the Eleventh International Symposium on Space Technology and Science* (Tokyo: Agne Publishing, 1975), 19.

29. Ibid., 35–36.

30. Kazuto Suzuki, personal communication with the author (December 18, 2010).

31. Berner, "Japan's Space Program," 3.

32. JAXA, "Timeline of ISAS."

33. Yamanouchi and Takenaka, "Space Development Activities in Japan," 62.

34. Berner, "Japan's Space Program," exhibit 2, "U.S. Firms Providing Technical Assistance, Production License, or Hardware for Japanese Launch Vehicles," 5.

35. Kazuto Suzuki, "Administrative Reforms and the Policy Logics of Japanese Space Policy," *Space Policy* 21, no. 1 (February 2005): 13.

36. Ibid.

37. Harvey, *The Japanese and Indian Space Programmes*, 22.

38. Matsura and Kuroda, "Progress Report on NASDA Projects," 36, and figure 1, p. 39.

39. JAXA, "NASDA History," http://www.jaxa.jp/about/history/nasda/index_e.html.

40. Harvey, *The Japanese and Indian Space Programmes*, 23.

41. Yamanouchi and Takenaka, "Space Development Activities in Japan," 54.

42. Pekkanen, *Picking Winners*, 174; Berner, "Japan's Space Program," 6.

43. Harvey, *The Japanese and Indian Space Programmes*, 29.

44. Ibid., 27.

45. Masao Yamanouchi and Yukihiko Takenaka, "Space Development Activities in Japan," in *Proceedings of the Fourteenth International Symposium on Space Technology and Science* (Tokyo: Agne Publishing, 1984), 53.

46. Harvey, *The Japanese and Indian Space Programmes*, 37–41.

47. JAXA, "NASDA History."

48. Yamanouchi and Takenaka, "Space Development Activities in Japan," 61.

49. On these events, see Harvey, *The Japanese and Indian Space Programmes*, 81–83.

50. Kazuto Suzuki, personal communication with the author (December 18, 2010).

51. David E. Sanger, "Japan Eyes Space with Uncertainty, and Confusion," *New York Times* (June 26, 1990): B5.

52. Ibid.

53. JAXA, "JAXA's Astronauts," http://www.jaxa.jp/projects/iss_human/astro/index_e.html.

54. David E. Sanger, "Japan Launches Rocket to the Moon," *New York Times* (January 25, 1990): A4.

55. Quoted in ibid.

56. Harvey, *The Japanese and Indian Space Programmes*, 42–43.

57. Ibid., 32.

58. Daikichiro Mori and Tamiya Nomura, "Japanese Observation Balloon, Sounding Rocket and Scientific Satellite Program," in *Proceedings of the Eleventh International Symposium on Space Technology and Science* (Tokyo: Agne Publishing, 1975), 14.

59. "Asia-Pacific Regional Space Agency Forum" (organizational brochure, acquired from the Japan Space Exploration Organization on April 22, 2009), undated.

60. Asia-Pacific Regional Space Agency Forum (APRSAF), "What's APRSAF?" http://www.aprsaf.org/about/.

61. Kazuto Suzuki, personal communication with the author (December 18, 2010).

62. Berner, "Japan's Space Program," 7.

63. Pekkanen, *Picking Winners*, 175.

64. Ibid.

65. Kazuto Suzuki, personal communication with the author (December 18, 2010).

66. Berner, "Japan's Space Program," 8.

67. Setsuko Aoki, "Current Status and Recent Developments in Japan's National Space Law and Its Relevance to Pacific Rim Space Law and Activities," *Journal of Space Law* 35, no. 2 (Winter 2009): 369.

68. Berner, "Japan's Space Program," 8.

69. Harvey, *The Japanese and Indian Space Programmes*, 59–60.

70. Kazuto Suzuki, "Japanese Steps Toward Regional and Global Confidence Building," in John M. Logsdon and James Clay Moltz, eds., *Collective Security in Space: Asian Perspectives* (Washington, D.C.: Space Policy Institute, George Washington University, January 2008), 135.

71. Quoted in Paul Kallender-Umezu, "Profile: Takeo Kawamura, Revamping Japan's Space Management," *Space News* (November 10, 2008): 18.

72. William W. Radcliffe, "Origins and Current State of Japan's Reconnaissance Satellite Program," *Studies in Intelligence* 54, no. 3 (September 2010): 11; and Berner, "Japan's Space Program," 17.

73. Berner, "Japan's Space Program," 18.

74. Ibid.

75. Ibid.

76. Suzuki, "Administrative Reforms and the Policy Logics of Japanese Space Policy," 16.

77. Quoted in Kallender-Umezu, "Profile: Takeo Kawamura," 18.

78. Ibid.

79. On these projects, see ibid., 97–100.

80. JAXA, "HTV (H-II Transfer Vehicle)," http://www.jaxa.jp/projects/rockets/htv/index_e.html.

81. JAXA, "Nozomi," http://www.isas.jaxa.jp/e/enterp/missions/nozomi/index.shtml.

82. JAXA, "Hayabusa's Return Journey to Earth: The Final Stage," http://www.jaxa.jp/article/special/hayabusa/index_e.html.

83. Tariq Mali, "Asteroid Dust Successfully Returned by Japanese Space Probe," *Space News* (November 22, 2010): 13.

84. JAXA, "Kaguya (Selene)," http://www.kaguya.jaxa.jp/index_e.htm.

85. JAXA, "Greenhouse gases Observing SATellite 'Ibuki' (GoSat)," http://www.jaxa.jp/projects/sat/gosat/index_e.html.

86. JAXA, "Kibo: Japanese Experimental Module," http://kibo.jaxa.jp/en/mission/.

87. Author's interview with Diet Member Katsuyuki Kawai, Tokyo, Japan (April 21, 2009).

88. Ibid.

89. Masakazu Toyoda, Secretary-General, Secretariat of Strategic Headquarters for Space Policy, Cabinet Secretariat, and handout entitled "Toward the Reform of Japanese Space Policy," page on "Contents of Basic Space Law," 3, provided to the author at the Cabinet Secretariat, Tokyo, Japan (April 2009).

90. Paul Kallender-Umezu, "Amid Power Shift in Power, Japan Seeks Space Budget Hike," *Space News* (September 7, 2009): 14.

91. Paul Kallender-Umezu, "Profile: Seiji Maehara, Foreign Minister, Japan," *Space News* (October 18, 2010): 22.

92. Quoted in Paul Kallender-Umezu, "Japan Urged to Break up JAXA and Establish New Space Agency," *Space News* (May 3, 2010): 10.

93. Kallender-Umezu, "Profile: Seiji Maehara."

94. Aoki, "Current Status and Recent Developments in Japan's National Space Law," appendix, p. 415, quoting the Basic Space Law.

95. "Mitsubishi Electric Lands ST-2 Satellite Contract," *Space News* (December, 8, 2008): 11.

96. Aoki, "Current Status and Recent Developments in Japan's National Space Law," 369.

97. Ibid.

98. Kazuto Suzuki, personal communication with the author (December 18, 2010).

99. Aoki, "Current Status and Recent Developments in Japan's National Space Law," 366.

100. "Japan's GX Rocket Targeted for Cancellation in 2010," *Space News* (November 23, 2009): 8.

101. "Undecided Japan Presses Ahead with SatNav Demo," *Space News* (March 16, 2009): 8; "Japanese Government Seeks To Reorient Space Spending," *Space News* (October 4, 2010): 7.

102. Ibid.

103. Kallender-Umezu, "Profile: Takeo Kawamura."

104. Author's interview with Japanese government official, Tokyo, Japan (April 22, 2009).

105. Author's interview with senior Japanese space official, Tokyo, Japan (April 20, 2009).

106. Ibid.

107. Watanabe, "Japan's National Space Strategy," 6.

108. Comment by industry representative at Keio University space forum (April 20, 2009).

109. Author's interviews with multiple senior Japanese space officials, Tokyo, Japan (April 20–23, 2009).

110. Ibid.

111. "Govt panel proposes early warning satellite," *Yomiuri Shimbun* (April 23, 2009).

112. "Japanese MoD Unveils First Ever Space Budget," *Space News* (September 8, 2009): 8.

113. "H-2A Rocket Lofts Japanese Reconnaissance Satellite," *Space News* (December 7, 2009).

114. Keisuke Yoshimura, "Japan's Spy Satellites Are an Open Secret," *Kyodo News* (June 15, 2007).

115. "Japan's Abe Charges China's Satellite Test Illegal," *Agence France Presse* (January 31, 2007). Japanese legal scholar Setsuko Aoki notes that Japan's stance on the Chinese ASAT drew from the International Law Association's draft debris treaty in 1994 in terms of its reliance on the Outer Space Treaty's Article IX as well as on Article XI on China's failure to notify others of its "scientific" program. Author's discussions with Setsuko Aoki, Tokyo, Japan (April 20, 2009).

116. Author's interview with senior Japanese official (name withheld), Tokyo, Japan (April 22, 2009).

117. Kazuto Suzuki, personal communication with the author (December 18, 2010).

118. Author's interviews with various Japanese officials, Tokyo, Japan (April 20–23, 2009).

119. Author's interview with senior Japanese official, Tokyo, Japan (April 20, 2009).

120. Pyle, *Japan Rising*, 360.

121. Yoshinobu Yamamoto, "Japan's Activism Lite: Bandwagoning the United States," in Byung-Kook Kim and Anthony Jones, eds., *Power and Security in Northeast Asia: Shifting Strategies* (Boulder, Colo.: Lynne Rienner, 2007), 128.

122. Bhubhindar Singh, "Japan's Security Policy: From a Peace State to an International State," *The Pacific Review* 21, no. 3 (July 2008): 306–308.

123. Asia-Pacific Regional Space Agency Forum, "APRSAF," organizational brochure, not dated, provided to author by JAXA personnel in Tokyo, Japan (April 22, 2009).

124. JAXA, "JPT Members," on the Sentinel Asia portion of the JAXA Web site, https://sentinel.tksc.jaxa.jp/sentinel2/MB_HTML/JPTMember/JPTMember.htm.

125. Author's interviews with Japanese space officials, Tokyo, Japan (April 20–23, 2009).

126. Author's interview with Diet Member Katsuyuki Kawai, Tokyo, Japan (April 21, 2009).

127. Author's interview with senior Japanese official, Tokyo, Japan (April 21, 2009).

128. News Briefs, "Six Asian Nations Partner on Satellite Development," *Space News* (June 8, 2009): 18.

129. JAXA, PowerPoint presentation, "JAXA Space Programs: Current and Future," given to author by JAXA personnel, Tokyo, Japan (April 22, 2009).

130. Ibid.

131. Author's interview with senior Japanese official, Tokyo, Japan (April 21, 2009).

132. Ibid.

133. Author's interview with Japanese space expert, Tokyo, Japan (April 20, 2009).

134. Pyle, *Japan Rising*, 373.

135. Pekkanen, *Picking Winners*, 184.

3. THE CHINESE SPACE PROGRAM: FROM TURBULENT PAST TO PROMISING FUTURE

1. See, for example, Gordon G. Chang, "The Space Arms Race Begins," *Forbes* (November 5, 2009).

2. Larry M. Wortzel, "The Chinese People's Liberation Army and Space Warfare," *Astropolitics* 6, no. 2 (May–August 2008): 124.

3. Joan Johnson-Freese, *Space as a Strategic Asset* (New York: Columbia University Press, 2007).

4. Fiona Cunningham, "The Stellar Status Symbol: True Motive for China's Manned Space Program," *China Security* 5, no. 3 (Winter 2009): 79, 71.

5. Kenneth Lieberthal, *Governing China: From Revolution Through Reform* (New York: W. W. Norton, 1995), 169.

6. Yeu-Farn Wang, *China's Science and Technology Policy: 1949–1989* (Aldershot: Avebury [Ashgate], 1993), 34.

7. Ibid., 35.

8. Roger Handberg and Zhen Li, *Chinese Space Policy: A Study in Domestic and International Politics* (New York: Routledge, 2007), 68–69.

9. For data on these individuals, see ibid., table 3.1, "Examples of returning scientists and technologists," 69.

10. Wang, *China's Science and Technology Policy*, 42; Handberg and Li, *Chinese Space Policy*, 60–61; and Joan Johnson-Freese, *The Chinese Space Program: A Mystery Within a Maze* (Malabar, Fla.: Krieger Publishing, 1998), 44.

11. Ibid.

12. Brian Harvey, *China's Space Program: From Conception to Manned Spaceflight* (Chichester: Praxis Publishing, 2004), 20–21.

13. Quoted in Cai Rupeng and Shao Xinfang, "China's Rocket Man," *NewsChina* (December 5, 2009): 37.

14. On this point, see Rosemary J. Foot, "Nuclear Coercion and the Ending of the Korean Conflict," *International Security* 13, no. 3 (Winter 1988/1989).

15. Johnson-Freese, *The Chinese Space Program*, 44–45.

16. Wang, *China's Science and Technology Policy*, 44–45; Handberg and Li, *Chinese Space Policy*, 62.

17. Johnson-Freese, *The Chinese Space Program*, 45.

18. Harvey, *China's Space Program*, 24.

19. Quoted Gregory Kulacki and Jeffrey G. Lewis, "A Place for One's Mat: China's Space Program: 1956–2003," American Academy of Arts & Sciences, Cambridge, Mass. (2009), 9.

20. Wang, *China's Science and Technology Policy*, 58.

21. Ibid., 6.

22. Richard P. Suttmeier, *Science, Technology, and China's Drive for Modernization* (Stanford, Calif.: Hoover Institution Press, 1980), 20–21.

23. Johnson-Freese, *The Chinese Space Program*, 45; Kulacki and Lewis, "A Place for One's Mat," 5; Wang, *China's Science and Technology Policy*, 57; and Harvey, *China's Space Program*, 27.

24. Harvey, *China's Space Program*, 29.

25. Ibid., 29–30; Kulacki and Lewis, "A Place for One's Mat," 8.

26. Wang, *China's Science and Technology Policy*, 59.

27. Ibid., 64.

28. Ibid., 65.

29. On the R-2's capabilities, see Asif A. Siddiqi, *Sputnik and the Soviet Space Challenge* (Gainesville: University of Florida Press, 2000), 97.

30. Harvey, *China's Space Program*, 34.

31. Ibid., 28.

32. Ibid., 37; Kulacki and Lewis, "A Place for One's Mat," 9.

33. Harvey, *China's Space Program*, 170.

34. Kulacki and Lewis, "A Place for One's Mat," 9–10.

35. Harvey, *China's Space Program*, 46.

36. For more on the Cultural Revolution's goals, see Lieberthal, *Governing China*, 111–113.

37. Document 16, "Red Guard Statements, 1966–67," "Declaring War on the Old World," Red Guards of Beijing No. 2 Middle School, in Gregor Benton and Alan Hunter, eds., *Wild Lily, Prairie Fire: China's Road to Democracy, Yan'an to Tian'anmen: 1942–1989* (Princeton, N.J.: Princeton University Press, 1995), 109.

38. Charles P. Ridley, *China's Scientific Policies: Implications for International Cooperation* (Washington, D.C.: American Enterprise, 1976), 12.

39. Wang, *China's Science and Technology Policy*, 72.

40. Ibid., 50–51.

41. Johnson-Freese, *The Chinese Space Program*, 48–49.

42. Harvey, *China's Space Program*, 49.

43. Handberg and Li, *Chinese Space Policy*, 74.

44. Harvey, *China's Space Program*, 52–53.

45. Ibid., 55.

46. Ibid., 60.

47. Handberg and Li, *Chinese Space Policy*, 74; Harvey, *China's Space Program*, 58–59.

48. Kulacki and Lewis, "A Place for One's Mat," 20.

49. Handberg and Li, *Chinese Space Policy*, 74. According to Harvey (*China's Space Program*, 241), this effort was named "Project 714" after the list was confirmed the following month (named for the year and month of final selection).

50. Johnson-Freese, *The Chinese Space Program*, 49; and Harvey, *China's Space Program*, 61–62.

51. Thomas Fingar, "Domestic Policy and the Quest for Independence," in Thomas Fingar, ed., *China's Quest for Independence in the 1970s* (Boulder, Colo.: Westview Press, 1980), 40.

52. On General Lin's death and the events that preceded his controversial flight, see Barbara Barnouin and Yu Changgen, *Ten Years of Turbulence: The Chinese Cultural Revolution* (London: Kegan Paul International, 1993), 222–246.

53. Handberg and Li, *Chinese Space Policy*, 78.

54. Ibid.

55. Kulacki and Lewis, "A Place for One's Mat," 20.

56. Ridley, *China's Scientific Policies*, table 7, "Chinese Medical and Scientific Delegations Abroad, 1973–74," 80.

57. Ibid., 84.

58. Johnson-Freese, *The Chinese Space Program*, 50.

59. Lieberthal notes, "Mao positioned Hua as a compromise candidate." See Lieberthal, *Governing China*, 123.

60. Suttmeier, *Science, Technology, and China's Drive for Modernization*, 1.

61. Johnson-Freese, *The Chinese Space Program*, 51.

62. Kulacki and Lewis, "A Place for One's Mat," 15.

63. Fingar, "Domestic Policy and the Quest for Independence," 71.

64. Suttmeier, *Science, Technology, and China's Drive for Modernization*, 51.

65. Ibid., 35.

66. Wang, *China's Science and Technology Policy*, 104.

67. Handberg and Li, *Chinese Space Policy*, 92.

68. Wang Chuan-shan, "Space Activities of Academia Sinica," in *Proceedings of the Thirteenth International Symposium on Space Technology and Science* (Tokyo: Agne Publishing, 1982), 58.

69. Suttmeier, *Science, Technology, and China's Drive for Modernization*, 78.

70. Kulacki and Lewis, "A Place for One's Mat," 16.

71. Harvey, *China's Space Program*, 191.

72. "Understanding on Cooperation in Space Technology Between the United States of America and the People's Republic of China" (not dated, but the document mentions that it follows the November–December 1978 visit), reproduced in Suttmeier, *Science, Technology, and China's Drive for Modernization*, 100–101.

73. Walter A. McDougall, "The Scramble for Space," *Wilson Quarterly* 4, no. 4 (Autumn 1980): 80.

74. Ibid., 245.

75. On the plan, see Suttmeier, *Science, Technology, and China's Drive for Modernization*, 2–4.

76. Ibid., 5.

77. Harvey, *China's Space Program*, 170.

78. Ridley, *China's Scientific Policies*, 78; Dinah Lee, "China Looks for Trade Liftoff," *Far Eastern Economic Review* (April 28, 1978): 14.

79. Handberg and Li, *Chinese Space Policy*, 69.

80. Kulacki and Lewis, "A Place for One's Mat," 17.

81. Ibid., 18; Johnson-Freese, *The Chinese Space Program*, 212; Handberg and Li, *Chinese Space Policy*, 86; Harvey, *China's Space Program*, 103.

82. Handberg and Li, *Chinese Space Policy*, 86.

83. Author's interview with Great Wall Industry Corporation official, Beijing, China (September 2009). The rest of the paragraph also draws on this interview and on related comments by other GWIC officials.

84. Ibid.

85. Harvey, *China's Space Program*, 117.

86. Kulacki and Lewis, "A Place for One's Mat," 21.

87. Wang, *China's Science and Technology Policy*, 157.

88. On this series of flights, see Harvey, *China's Space Program*, 86–87.

89. Kulacki and Lewis, "A Place for One's Mat," 23.

90. Ibid.

91. Ibid., 24.

92. Nuclear Threat Initiative, "State Administration for Science, Technology and Industry for National Defense (SASTIND)," NTI Research Library, http://www.nti.org/db/china/costind.htm.

93. Author's interviews in Beijing, China (September 2009); see also Gregory Kulacki and Jeffrey G. Lewis, "Understanding China's ASAT Test," *Nonproliferation Review* 15, no. 2 (July 2008).

94. James A. Lewis, "China as a Military Space Competitor," in John M. Logsdon and Audrey M. Schaffer, eds., *Perspectives on Space Security* (Washington, D.C.: Space Policy Institute, George Washington University, December 2005), 96.

95. Dayao Li, Shaochun Chi, and Shiju Jiao, "China's Satellite Remote Sensing Technology and Its Application in [the] 20th Century," paper presented at the Twenty-second Asian Conference on Remote Sensing, Singapore (November 5–9, 2001).

96. Harvey, *China's Space Program*, 248.

97. Ibid.

98. Ibid., 248–249.

99. Tai Ming Cheung, *Fortifying China: The Struggle to Build a Modern Defense Economy* (Ithaca, N.Y.: Cornell University Press, 2009), 253.

100. China Aerospace Science and Technology Corporation (CASC), "Company Profile," on the CASC Web site, http://www.spacechina.com/english/about_01.shtml.

101. Ibid.

102. Harvey, *China's Space Program*, 113–114.

103. Author's interview with Great Wall Industry Corporation official, Beijing, China (September 2009).

104. Harvey, *China's Space Program*, 116.

105. Ibid., 117–118, 327.

106. Ibid., 121; Handberg and Li, *Chinese Space Policy*, 107; Johnson-Freese, *The Chinese Space Program*, 80. Figures for the injured range from twenty-three to fifty-six people.

107. Johnson-Freese, *The Chinese Space Program*, 80–81; Harvey, *China's Space Program*, 122.

108. Johnson-Freese, *The Chinese Space Program*, 86.

109. U.S. House of Representatives, "Report of the Select Committee on U.S. National Security and Military/Commercial Concerns with the People's Republic of China" (unclassified version), Report 105–851 (May 1999), http://www.access.gpo.gov/congress/house/hr105851/index.html.

110. See Alastair Iain Johnston, W. K. H. Panofsky, Marco Di Capua, and Lewis R. Franklin, *The Cox Committee Report: An Assessment*, ed. M. M. May (Stanford, Calif.: Center for International Security and Cooperation, Stanford University, December 1999).

111. Johnson-Freese, *The Chinese Space Program*, 102–103.

112. Harvey, *China's Space Program*, 160–161.

113. APSCO, "History," http://www.apsco.int/history.aspx. The rest of this paragraph draws on this official history of the organization.

114. Hu Xiaodi, Ambassador for Disarmament Affairs, Head of Chinese CD Delegation, "China's Position on and Suggestions for Ways to Address the Issue of Prevention of an Arms Race in Outer Space at the Conference on Disarmament," Working Paper, CD/1606 (February 9, 2000), 6, 4.

115. Hu Xiaodi, Ambassador for Disarmament Affairs, Head of Chinese CD Delegation, "Possible Elements of the Future International Legal Instrument on the Prevention of the Weaponization of Outer Space," Working Paper, CD/1645 (June 6, 2001).

116. Ibid., 3.

117. Hu Xiaodi, Ambassador for Disarmament Affairs, Head of Chinese CD Delegation, and Leonid A. Skotnikov, Ambassador, Permanent Representative of the Russian Federation to the CD, "Possible Elements for a Future International Legal Agreement on the Prevention of the Deployment of Weapons in Outer Space, the Threat or Use of Force Against Outer Space Objects," Working Paper, CD/1679 (June 22, 2002).

118. Johnson-Freese, *Space as a Strategic Asset*, 10.

119. Leonard David, "Shenzhou-7 Deployed Picosat That Came Within 25 Kilometers of Space Station," *Space News* (October 20, 2008): 10.

120. Chinese State Council, "China's Space Activities in 2006."

121. "China Sends Chang'e Probe Crashing Into Moon's Surface," *Space News* (March 9, 2009): 9.

122. Associated Press, "Russia, China Plan Joint Space Projects" (November 9, 2006).

123. AP-MCSTA Secretariat, "The Inauguration Ceremony of APSCO Held in Beijing," *Asia-Pacific Space Outlook* 4, no. 19 (December 2008): 2.

124. Chinese National Space Administration, "HJ-A/B of Environment and Disasters Monitoring Microsatellite Constellation Delivered to the Users," http://www.cnsa.gov.cn/n615709/n620682/n639462/168207.html.

125. Author's interview with State Department official involved in the Griffin trip [name withheld] (April 2009).

126. See Warren Ferster and Amy Klamper, "China, U.S. Put Spaceflight Cooperation Talks on Agenda," *Space News* (November 23, 2009): 12.

127. Quoted in Amy Klamper, "Human Spaceflight on Agenda for Bolden's China Trip," *Space News* (October 11, 2010): 14.

128. Quoted in Amy Svitak, "Bolden Details Trip to China During Marshall Visit," *Space News* (November 29, 2010): 11.

129. Rui C. Barbosa, "China Launches Military Satellite YaoGan Weixing-10," NASAspaceflight.com (August 9, 2010), http://www.nasaspaceflight.com/2010/08/china-launches-military-satellite-yaogan-weixing-10/.

130. On the test, see Bharath Gopalaswamy and Subrata Ghoshroy, "Is Space Weaponization Inevitable?" *Space News* (May 5, 2008): 19; Joan Johnson-Freese, *Heavenly Ambitions: America's Quest to Dominate Space* (Philadelphia: University of Pennsylvania Press, 2009), 10–11.

131. J.-C. Liou and N. L. Johnson, "Physical Properties of the Large Fengyun-1 Breakup Fragments," *Orbital Debris Quarterly* 12, no. 2 (April 2008): 4.

132. "Chinese Anti-satellite Test Creates Most Severe Orbital Debris Cloud in History," *Orbital Debris Quarterly* 11, no. 2 (April 2007): 2.

133. Kulacki and Lewis, "Understanding China's ASAT Test," 344.

134. Chinese State Council, "China's Space Activities in 2006," posted on the Web site of the Chinese National Space Administration, http://www.cnsa.gov.cn/n615709/n620681/n771967/79970.html.

135. On this issue, see You Ji and Daniel Alderman, "Changing Civil-Military Relations in China," in Roy Kamphausen, David Lai, and Andrew Scobell, eds., *The PLA at Home and Abroad: Assessing the Operational Capabilities of the Chinese Military* (Carlisle, Penn.: Army War College, 2010), 148–149.

136. Comment by Chinese participant at the conference on "China, Space, and Strategy III," University of British Columbia, Vancouver, Canada (September 4, 2008).

137. Valery Loshchinin, Ambassador and Permanent Representative of the Russian Federation to the CD, and Wang Qun, Ambassador of Disarmament Affairs and Head of Chinese CD Delegation, "Treaty on the Prevention of the Placement of Weapons in

Outer Space and of the Threat or Use of Force Against Outer Space Objects (PPWT)," Draft Treaty, CD/1839 (February 29, 2008).

138. Ibid., 3.

139. Wang Qun, Ambassador of Disarmament Affairs and Head of Chinese CD Delegation, and Valery Loshchinin, Ambassador and Permanent Representative of the Russian Federation to the CD, Answers to the Principal Questions and Comments to the Draft "Treaty on the Prevention of the Placement of Weapons in Outer Space and of the Threat or Use of Force Against Outer Space Objects (PPWT)," Annex, CD/1872.

140. Julian E. Barnes and Jeremy Page, "China Snubs U.S. Defense Pitch," *Wall Street Journal* (January 11, 2011): A9.

141. "China's Space Activities (White Paper)."

142. Ibid.

143. Andy Pasztor, "Satellite Firm Prepares for Launches by China," *Wall Street Journal* (April 28, 2008): B3.

144. Peter B. De Selding, "Eutelsat Unapologetic About Use of Chinese Launch Services," *Space News* (March 16, 2009): 1.

145. Andy Pasztor, "China to Launch Satellite for France's Eutelsat," *Wall Street Journal* (February 25, 2009): A5.

146. On this issue, see Rep. Dana Rohrabacher, "Dangers of Chinese Satellite Technology," *Space News* (April 13, 2009): 19.

147. Peter B. de Selding, "SES, Intelsat Asking Lawmakers to Rethink Launch Ban on China, India," *Space News* (August 3, 2009): 1.

148. Kevin Pollpeter, "Building for the Future: China's Progress in Space Technology During the Tenth 5-Year Plan and the U.S. Response," U.S. Army War College, Strategic Studies Institute, March 2008, p. 13.

149. Ibid.

150. Peter B. de Selding, "European Officials Poised to Remove Chinese Payloads from Galileo Sats," *Space News* (March 12, 2010), http://www.spacenews.com/policy/100312-officials-poised-remove-chinese-payloads-galileo.html.

151. Andrew Browne, "Beijing Says Billions in Funds Are Missing," *Wall Street Journal* (December 30, 2009): A9.

152. Susan L. Shirk, *China: Fragile Superpower* (New York: Oxford University Press, 2008), 6.

153. Rui C. Barbosa, "China Opens 2010 with BeiDou2 Satellite Launch," NASAs paceflight.com (January 16, 2010), http://www.nasaspaceflight.com/2010/01/china-opens-2010-with-beidou-2-satellite-launch/.

154. Peter B. de Selding, "China Satnav System Planned for 2010; Frequency Issue Still Unresolved," *Space News* (May 5, 2008): 7.

155. Quoted in de Selding, "China Satnav System Planned for 2010."

156. "70 Successful Launches in 2010," *New Papyrus* (January 3, 2011), http://new papyrusmagazine.blogspot.com/2011/01/70-successful-space-launches-in-2010.html.

157. Author's interview with Great Wall Industry Corporation official, Beijing, China (September 2009).

158. Xinhua News Agency, "China's New Carrier Rocket to Debut in 2014" (March 2, 2008).

159. Bradley Perrett, *Aerospace Daily and Defense Report* (March 6, 2008).

160. Peter J. Brown, "China Needs Sharper Eyes in Space," *Asia Times* (October 16, 2008), http://www.atimes.com/atimes/China/JJ16Ad02.html.

161. Chinese State Council, "China's Space Activities in 2006."

SEQ BNRef * MERGEFORMAT 162. "China's Space Activities (White Paper)."

163. Handberg and Li, *Chinese Space Policy*, 172.

164. Li Juqian, "Progressing Towards New National Space Law: Current Legal Status and Recent Developments in Chinese Space Law and Its Relevance to Pacific Rim Space Law and Activities," *Journal of Space Law* 35, no. 2 (Winter 2009): 444.

165. Li Shouping, "The Role of International Law in Chinese Space Law and Its Relevance to Pacific Space Law and Activities," *Journal of Space Law* 35, no. 2 (Winter 2009): 555.

166. Eric Hagt, "Emerging Grand Strategy for China's Defense Industry Reform," in Roy Kamphausen, David Lai, and Andrew Scobell, eds., *The PLA at Home and Abroad: Assessing the Operational Capabilities of the Chinese Military* (Carlisle, Penn.: Army War College, 2010), 514.

167. Ibid., 484.

168. Nuclear Threat Initiative, "State Administration for Science, Technology and Industry for National Defense (SASTIND)"; James Mulvenon and Rebecca Samm Tyroler-Cooper, "China's Defense Industry on the Path of Reform," report prepared for the U.S.-China Economic and Security Review Commission (October 2009).

169. Ibid.; also Federation of American Scientists, "Chinese Aerospace Science and Technology Corporation (CASC), Chinese Aerospace Corporation (CASC) [*sic*]," on the FAS Web site, http://www.fas.org/nuke/guide/china/contractor/casc.htm.

170. China Aerospace Science and Technology Corporation (CASC), "About CASC," on the CASC Web site, http://www.spacechina.com/english/about_01.shtml.

171. Mulvenon and Tyroler-Cooper, "China's Defense Industry on the Path of Reform," 19.

172. Pollpeter, "Building for the Future," viii.

173. Edward Cody, "China Builds and Launches a Satellite for Nigeria," *Washington Post* (May 14, 2007): 11.

174. Peter J. Brown, "Chavez Cherishes His Chinese-Built Satellite," *Asia Times* (August 6, 2009), http://www.atimes.com/atimes/China/KH06Ad02.html.

175. "China to Build Satellite for Pakistan," *Space News* (October 20, 2008): 3.

176. Great Wall Industry Corporation, "LaoSat-1 Program," on the GWIC Web site, http://www.cgwic.com.cn/In-OrbitDelivery/CommunicationsSatellite/Program/Laos.html.

177. "China to Build and Launch Bolivian Telecom Satellite," *Space News* (April 5, 2010): 3.

178. "Intelsat, China Central TV Extend Services Contract," *Space News* (June 23, 2008): 9.

179. Jason Dean, "China's Xinhua to Launch English-Language Channel," *Wall Street Journal* (May 1–2, 2010): A14.

180. Shouping, "The Role of International Law in Chinese Space Law," 543.

181. On this issue, see Stephanie Lieggi, "From Proliferator to Model Citizen? China's Recent Enforcement of Nonproliferation-Related Trade Controls and Its Potential Positive Impact in the Region," *Strategic Studies Quarterly* 4, no. 2 (Summer 2010).

182. "China Plans to Launch Space Station by 2011," *Space News* (October 6, 2008): 9.

183. Ibid.; "China Sends Chang'e Probe Crashing Into Moon's Surface."

184. Amy Klamper, "Official Details 11-Year Path to Developing China's Own Space Station," *Space News* (April 14, 2010), http://www.spacenews.com/civil/100414 -path-china-space-station.html.

185. Ibid.

186. Space Foundation, *The Space Report: 2010* (Colorado Springs: Space Foundation, 2010), "Chinese Human Spaceflight Program," 115.

187. Yi Zhou, "Perspectives on Sino-US Cooperation in Civil Space Programs," *Space Policy* 24, no. 3 (August 2008): 136–137.

188. Ibid., 139.

189. Handberg and Li, *Chinese Space Policy*, 115.

190. Cartwright, cited in Johnson-Freese, *Heavenly Ambitions*, 11; also quoted in Elaine M. Grossman, "Is China Disrupting U.S. Satellites?" *InsideDefense.com* (October 13, 2006), http://www.military.com/features/0,15240,116694,00.html?ESRC=topstories.RSS.

191. On this point, see Lewis, "China as a Military Space Competitor," 93.

192. On these points, see ibid.; see also Jeffrey Lewis, *The Minimum Means of Reprisal: China's Search for Security in the Nuclear Age* (Cambridge, Mass.: The MIT Press, 2007).

193. Ibid., 105.

194. Xiaoming Zhang and Col. Sean D. McClung, "The Art of Military Discovery: Chinese Air and Space Power Implications for the USAF," *Strategic Studies Quarterly* 4, no. 1 (Spring 2010): 47.

195. Wortzel, "The Chinese People's Liberation Army and Space Warfare," 114.

196. Handberg and Li, *Chinese Space Policy*, 173.

197. On this point, see ibid.

198. Pollpeter, "Building for the Future," viii.

199. Wortzel, "The Chinese People's Liberation Army and Space Warfare," 113.

200. Lieggi, "From Proliferator to Model Citizen?" 58–59.

4. THE INDIAN SPACE PROGRAM: RISING TO A CHALLENGE

1. News Briefs, "India Developing Means to Destroy Satellites," *Space News* (January 11, 2010): 9.

2. Brian Harvey, *The Japanese and Indian Space Programmes: Two Roads Into Space* (London: Springer, 2000), 127.

3. David Arnold, *Science, Technology, and Medicine in Colonial India* (Cambridge: Cambridge University Press, 2000).

4. N. Pant, "Space Activities in India," in *Proceedings of the Fourteenth International Symposium on Space Technology and Science* (Tokyo: Agne Publishing, 1984), 21.

5. Jean-Paul Escallettes and Philippe Jung, "William Congreve and the City of Toulouse," in John Becklake, ed., *History of Rocketry and Astronautics*, American Astronautical Society History Series 17 (San Diego, Calif.: Univelt, 1995), 14.

6. K. K. Nair, *Space: The Frontiers of Modern Defence* (New Delhi: Knowledge World, 2006), 2.

7. Arnold, *Science, Technology, and Medicine in Colonial India*, 186.

8. Ibid., 169.

9. Ibid., 196.

10. Harvey, *The Japanese and Indian Space Programmes*, 128.

11. Ibid.

12. Arnold, *Science, Technology, and Medicine in Colonial India*, 207.

13. Ibid., 210.

14. Jawaharlal Nehru, "Science & Technology," in *Indian Perspectives* (New Delhi) (August 2008): 42.

15. On the Nehru government's economic policies, see Ranbir Vohra, *The Making of India: A Historical Survey* (Armonk, N.Y.: M. E. Sharpe, 1997), 211–215.

16. S. K. Kumar, "Indian Astronautical Society: A Brief Report on Its Working, History, and Development," in F. Tamaki, ed., *Proceedings of the Second International Symposium on Rockets and Astronautics* (Tokyo: Yokendo, 1961), 328.

17. Pant, "Space Activities in India," 21; Department of Space, *Annual Report: 2006–2007*, 93.

18. Sarabhai, quoted in B. N. Suresh, "History of Indian Launchers," *Acta Astronautica* 63 (2008): 428.

19. S. Dhawan and U. R. Rao, "The Indian Space Programme," in *Proceedings of the Thirteenth International Symposium on Space Technology and Science* (Tokyo: Agne Publishing, 1982), 22.

20. Sundara Vadlamudi, "Indo-U.S. Space Cooperation: Poised for Take-Off?" *The Nonproliferation Review* 12, no. 1 (March 2005), table 1, "Indo-U.S. space cooperation," p. 203.

21. Suresh, "History of Indian Launchers," 429.

22. Percival Spear, *India: A Modern History* (Ann Arbor: University of Michigan Press, 1961), 444.

23. On this point, see Harvey, *The Japanese and Indian Space Programmes*, 129.

24. G. B. Pant, "Trip to Moon" [*sic*], *Proceedings of the First All-India Symposium on Rocket Technology* (Bihar: Ranchi University, 1967), 272.

25. Rajeev Lochan, "Some Reflections on Collective Security in Space," in John M. Logsdon and James Clay Moltz, eds., *Collective Security in Space: Asian Perspectives* (Washington, D.C.: George Washington University, Space Policy Institute, January 2008), 33.

26. Harvey, *The Japanese and Indian Space Programmes*, 130.

27. Dinshaw Mistry, "The Geostrategic Implications of India's Space Program," *Asian Survey* 41, no. 6 (November/December 2001): 1034; Dhawan and Rao, "The Indian Space Programme," 22.

28. Dhawan and Rao, "The Indian Space Programme," 22.

29. Harvey, *The Japanese and Indian Space Programmes*, 136.

30. V. S. Mani, "Space Policy and Law in India and Its Relevance to the Pacific Rim," *Journal of Space Law* 35, no. 2 (Winter 2009): 618–619.

31. Ibid., 619.

32. Dhawan and Rao, "The Indian Space Programme," 23.

33. Ibid.

34. Daphne Burleson, "Indian Space Research Organization (ISRO)," in her *Space Programs Outside the United States: All Exploration and Research Efforts, Country by Country* (Jefferson, N.C.: McFarland & Company, 2005), 144.

35. Vadlamudi, "Indo-U.S. Space Cooperation: Poised for Take-Off?" table 1, "Indo-U.S. space cooperation," p. 203.

36. Ibid., 202.

37. Ibid., 200.

38. Harvey, *The Japanese and Indian Space Programmes*, 134.

39. Ibid., 135.

40. Ibid.

41. Ibid.

42. Ibid., 136.

43. Dhawan and Rao, "The Indian Space Programme," figure 2, "ISRO Satellite Missions," p. 29.

44. Ibid., 22.

45. Ibid., 23–24; Department of Space, *Annual Report: 2006–2007*, 93.

46. Dhawan and Rao, "The Indian Space Programme," 23.

47. Vadlamudi, "Indo-U.S. Space Cooperation: Poised for Take-Off?" 202.

48. Dhawan and Rao, "The Indian Space Programme," 24.

49. Dennis Kux, "India at Sixty: A Positive Balance Sheet," *Headline Series*, no. 330, Foreign Policy Association (Fall 2007): 42.

50. Harvey, *The Japanese and Indian Space Programmes*, 163.

51. Ibid., 165.

52. Mistry, "The Geostrategic Implications of India's Space Program," 1026.

53. Vadlamudi, "Indo-U.S. Space Cooperation: Poised for Take-Off?" 206.

54. Mistry, "The Geostrategic Implications of India's Space Program," 1029.

55. Ibid., 1029, 1034.

56. Indian Space Research Organisation (ISRO), "Launch Vehicles," http://www.isro.org/Launchvehicles/launchvehicles.aspx#ASLV (accessed May 7, 2010).

57. Mistry, "The Geostrategic Implications of India's Space Program," table 1, "India's Satellite Launch Vehicles and Missiles," p. 1027.

58. Suresh, "History of Indian Launchers," 431.

59. Vadlamudi, "Indo-U.S. Space Cooperation: Poised for Take-Off?" 206.

60. Pant, "Space Activities in India," 27–28.

61. Burleson, "Indian Space Research Organization (ISRO)," 146.

62. Vohra, *The Making of India*, 277.

63. Ibid., 278.

64. See the Web site of the Antrix Corporation Limited, http://www.antrix.gov.in/aboutus.html.

65. ISRO, "Polar Satellite Launch Vehicle," undated pamphlet from ISRO, Bangalore, India, section on "About PSLV."

66. Ibid., see chart, "Previous PSLV Launches."

67. Suresh, "History of Indian Launchers," 432.

68. Department of Space, *Annual Report: 2006–2007*, "Space Transportation," 48.

69. Susan Eisenhower, et al., *Partners in Space: U.S.-Russian Cooperation After the Cold War* (Washington, D.C.: Eisenhower Institute, 2004), 33; Nuclear Threat Initiative (NTI), "Russia: Missile Exports to India: Components" (January 13, 1999), http://www.nti.org/db/nisprofs/russia/exports/rusind/comp.htm.

70. George Perkovich, *India's Nuclear Bomb: The Impact on Global Proliferation* (Berkeley: University of California Press, 1990), 248.

71. Richard Speier, "U.S. Satellite Space Launch Cooperation and India's Intercontinental Ballistic Missile Program," in Henry Sokolski, ed., *Gauging U.S.-Indian Strategic Cooperation* (Carlisle, Penn.: U.S. Army War College, Strategic Studies Institute, March 2007), 194.

72. Suresh, "History of Indian Launchers," 433.

73. Kux, "India at Sixty," 45.

74. On these agreements, see Vadlamudi, "Indo-U.S. Space Cooperation: Poised for Take-Off?" 205–206.

75. Mistry, "The Geostrategic Implications of India's Space Program," 1039.

76. Nair, *Space: The Frontiers of Modern Defence*, 17.

77. Ibid.

78. Vadlamudi, "Indo-U.S. Space Cooperation: Poised for Take-Off?" 207.

79. Ibid., 207–208.

80. Teresita C. Schaffer, "Partnering with India: Regional Power, Global Hopes," in Ashley J. Tellis, Mercy Kuo, and Andrew Marble, eds., *Strategic Asia, 2008–09: Challenges and Choices* (Seattle, Wash.: National Bureau of Asian Research, 2008), 214.

81. Kux, "India at Sixty," 58.

82. Pallava Bagla, "US 'Red Tape' Dogged India Moon Mission," *BBC News* (October 1, 2009), http://news.bbc.co.uk/2/hi/south_asia/8281480.stm.

83. "NASA Instrument on Candrayaan-1 Finds Water on Moon," *Space News* (September 28, 2009): 8.

84. Kux, "India at Sixty," 45.

85. On this point, see Anand Giridharadas, "India Calling: Their Parents Came to America, but These Days the Motherland Beckons," *New York Times* (November 23, 2008).

86. Suresh, "History of Indian Launchers," 432.

87. Bharath Gopalaswamy and Subrata Ghoshroy, "Is Space Weaponization Inevitable?" *Space News* (May 5, 2008): 19; Schaffer, "Partnering with India," 443.

88. K. S. Jayaraman, "India's Space Cell Leverages ISRO Technology for Armed Forces," *Space News* (February 16, 2009): A4.

89. Space Foundation, *The Space Report: 2010* (Colorado Springs: Space Foundation, 2010), 68.

90. Peter J. Brown, "India's Space Program Takes a Hit," *Asia Times* (May 1, 2010), http://www.atimes.com/atimes/South_Asia/LE01Df01.html.

91. Department of Space, *Annual Report: 2006–2007*, "Space Industry Partnership," 73.

92. Becky Iannotta, "Satellite TV in India Expected to Triple by 2013," *Space News* (October 27, 2008): 10.

93. Lochan, "Some Reflections on Collective Security in Space," 34.

94. Ibid.

95. Ibid.

96. Mistry, "The Geostrategic Implications of India's Space Program," 1042.

97. Somini Sengupta, "As Indian Growth Soars, Child Hunger Persists," *New York Times* (March 13, 2009): A1.

98. Mistry, "The Geostrategic Implications of India's Space Program," 1042.

99. Bharath Gopalalswamy, "Indian Space Policy: Aiming Higher," *Space News* (July 21, 2008): 23.

100. Krishnaswami Kasturirangan, "Indian Space Programme," *Acta Astronautica* 54 (2004): 843.

101. News Briefs, "Six Asian Nations Partner on Satellite Development," *Space News* (June 8, 2009): 18.

102. Brown, "India's Space Program Takes a Hit."

103. Ajey Lele, "Space Security and India," in Sukhvinder Kaur Multani, ed., *Space Security* (Hyderabad: Icfai University Press, 2008), 183.

104. Ibid.

105. Ibid.

106. Author's interview with Indian space official, Washington, D.C. (November 20, 2008).

107. Geeta Anand, "India's Colleges Battle a Thicket of Red Tape," *Wall Street Journal* (November 13, 2008): A1.

108. On this point, see Nair, *Space: The Frontiers of Modern Defence*, 174.

109. Ibid., 250.

110. Space Foundation, *The Space Report: 2010*, 122.

111. Dipankar Banerjee, "Indian Perspectives on Regional Space Security," in John M. Logsdon and James Clay Moltz, eds., *Collective Security in Space: Asian Perspectives* (Washington, D.C.: George Washington University, Space Policy Institute, January 2008), 129.

112. Nair, *Space: The Frontiers of Modern Defence*, 177.

113. Ibid.

114. Ibid.

115. Jayaraman, "India's Space Cell Leverages ISRO Technology for Armed Forces."

116. "India Launches Its First Radar Satellite," *Space News* (April 27, 2009): 9.

117. "ISRO Releases First Images from Cartosat-2B Satellite," *Space News* (July 26, 2010): 8.

118. Jayaraman, "India's Space Cell Leverages ISRO Technology for Armed Forces."

119. Ibid.

120. Space Foundation, *The Space Report: 2010*, 122.

121. Ibid.

122. Ibid.

123. Mani, "Space Policy and Law in India," 630.

124. Ibid., 631.

125. Ibid., 630.

126. Lele, "Space Security and India," 185.

127. Nair, *Space: The Frontiers of Modern Defence*, 133.

128. V. K. Saraswat, quoted in Peter J. Brown, "India Targets China's Satellites," *Asia Times* (January 22, 2010), http://www.atimes.com/atimes/South_Asia/LA22Df01.html.

129. Naik, quoted in Victoria Samson, "India's Missile Defense/Anti-satellite Nexus," *Space Review* (May 10, 2010), http://www.thespacereview.com/article/1621/1.

130. Saraswat, quoted in Indira Bagchi, "India Working on Tech to Defend Satellites," *Times of India* (March 6, 2011).

131. Ibid.

132. Sonya Misquitta, "Defense Contractors Target Big Jump in India's Military Spending," *Wall Street Journal* (July 17, 2009): B1, citing figures for India's 2010 fiscal year.

133. "India Adding Space Radar to Missile Shield by 2014," *Space News* (March 16, 2009): 9.

134. Nair, *Space: The Frontiers of Modern Defence*, 178.

135. M. P. Anil Kumar, "Let Us Develop a Military Space Programme," Rediff.com (July 10, 2009).

136. Former DRDO chief Aatre, quoted in Radhakrishna Rao, "Chinese Threat to Indian Space Assets," *domain-b.com* (January 29, 2009), http://www.domain-b.com/aero/20090129_indian_space.html.

137. News Briefs, "Anti-satellite Weapons Part of Indian Technology Vision," *Space News* (June 7, 2010): 8.

138. D. Narayana Moorthi, "What 'Space Security' Means to an Emerging Space Power," *Astropolitics* 2, no. 2 (Summer 2004): 263.

139. On the Indian perspective, see M. Y. S. Prasad, "Technical and Legal Issues Surrounding Space Debris—India's Position in the UN," *Space Policy* 21, no. 4 (November 2005).

140. Author's interviews with participants in the debris-mitigation discussions during 2006 to 2007.

141. Jitendra Nath Goswami and Mylswamy Annadurai, "Chandrayaan-1 Mission to the Moon," *Acta Astronautica* 63 (2008): 1216; Vibhuti Agarwal, "India to Launch Its First Unmanned Moon Mission," *Wall Street Journal* (October 20, 2008): A13.

142. Peter Wonacott, "India Befriends Afghanistan, Irking Pakistan," *Wall Street Journal* (August 19, 2009): A8.

143. K. S. Jayaraman, "India Plans First Manned Mission with Assistance from Russian Space Agency," *Space News* (February 2, 2009): 11; Space Foundation, *The Space Report: 2010*, 47.

144. Jayaraman, "India Plans First Manned Mission with Assistance from Russian Space Agency."

145. Space Commission member and senior ISRO official K. Radharishnan, quoted in "India Touts Plans to Hoist Tricolour on Moon By 2020," *Moon Daily* (January 5, 2009), http://www.moondaily.com/reports/India_Touts_Plans_To_Hoist_Tricolour_On_Moon_By_2020_999.html.

146. Space analyst Asif Siddiqi, quoted in Brown, "India's Space Program Takes a Hit."

147. Anatoly Zak, "Cooperation with India," http://www.russianspaceweb.com/luna_resurs.html.

148. Max Martin, "Over the Moon India Set for Mars Mission," *India Today* (January 28, 2010), http://indiatoday.intoday.in/site/Story/81262/Over+the+moon+India+set+for+Mars+mission.html.

149. "United States, India Pledge Expanded Civil Space Ties," *Space News* (November 15, 2010): 9.

150. Schaffer, "Partnering with India," 219.

151. Ibid.

152. Vadlamudi, "Indo-U.S. Space Cooperation," 216.

153. Evan A. Feigenbaum, "India's Rise, America's Interest," *Foreign Affairs* 89, no. 2 (March/April 2010): 86.

154. Ibid. In his essay, Feigenbaum only applies this insightful term to economic issues, but it can be equally applied to military and political affairs.

155. Arpan Mukherjee and Abhrajit Gangopadhyay, "India, China Aim to Double Trade," *Wall Street Journal* (December 17, 2010): A15.

156. Editorial, "India's Anti-satellite Comments Illustrate Underlying Reality," *Space News* (January 18, 2010): 18.

5. THE SOUTH KOREAN SPACE PROGRAM: EMERGING FROM DEPENDENCY

1. Doo Hwan Kim, "Korea's Space Development Programme: Policy and Law," *Space Policy* 22, no. 1 (February 2006): 110.

2. On the use of this term in a general political/alliance context within Asia, see Woosang Kim, "Korea as a Middle Power in the Northeast Asian Security Environ-

ment," in G. John Ikenberry and Chung-in Moon, *The United States and Northeast Asia: Debates, Issues, and New Order* (Lanham, Md.: Rowman and Littlefield, 2008).

3. Wade L. Huntley, "Smaller State Perspectives on the Future of Space Governance," *Astropolitics* 5, no. 3 (September–December 2007): 253.

4. Ibid., 254.

5. Ibid.

6. Changdon Kee, "A South Korean Perspective on Strengthening Space Security in East Asia," in John M. Logsdon and James Clay Moltz, eds., *Collective Security in Space: Asian Perspectives* (Washington, D.C.: George Washington University, Space Policy Institute, January 2008), 16.

7. Kyung-Min Kim, "South Korean Capabilities for Space Security," in John M. Logsdon and James Clay Moltz, eds., *Collective Security in Space: Asian Perspectives* (Washington, D.C.: George Washington University, Space Policy Institute, January 2008), 74.

8. Young-Gul Kim, "Innovation and the Role of Korea's Universities," in Lewis M. Branscomb and Young-Hwan Choi, eds., *Korea at the Turning Point: Innovation-Based Strategies for Development* (Westport, Conn.: Praeger, 1996), 126.

9. Uk Heo, Houngcheul Jeon, Hayam Kim, and Okjin Kim, "The Political Economy of South Korea: Economic Growth, Democratization, and Financial Crisis," *Maryland Series in Contemporary Asian Studies* 2, no. 193 (2008): 2.

10. Young-Hwan Choi, "The Path to Modernization: 1962–1992," in Lewis M. Branscomb and Young-Hwan Choi, eds., *Korea at the Turning Point: Innovation-Based Strategies for Development* (Westport, Conn.: Praeger, 1996), 13.

11. Carter J. Eckert, Ki-baik Lee, Young Ick Lew, Michael Robinson, and Edward W. Wagner, *Korea Old and New: A History* (Seoul: Ilchokak Publishers, 1990), 361.

12. Young-Gul Kim, "Innovation and the Role of Korea's Universities," 127.

13. Ibid.

14. Ibid.

15. Daniel A. Pinkston, "North and South Korean Space Development: Prospects for Cooperation and Conflict," *Astropolitics* 4, no. 2 (Summer 2006): 210.

16. Ibid., 212.

17. Wyn Bowen argues that the ROK also developed some 220–250 km.-range surface-to-surface missiles as part of this program. See Wyn Bowen, *The Politics of Ballistic Missile Proliferation* (London: MacMillan Press, 2000), 34.

18. Daniel A. Pinkston, "Space Cadets: The Korean Peninsula's Rocket Competition," *Jane's Intelligence Review* (September 2009): 9.

19. Ibid.

20. Young-Gul Kim, "Innovation and the Role of Korea's Universities," 128.

21. J. L. Enos and W.-H. Park, *The Adoption and Diffusion of Imported Technology: The Case of Korea* (New York: Croom Helm, 1988), 39.

22. Ibid., 244.

23. Kee, "A South Korean Perspective on Strengthening Space Security in East Asia," 14. Also, Pinkston, "North and South Korean Space Development," 211.

24. In 1990, the Roh government also normalized relations with the Soviet Union, setting the stage for later space cooperation with the Russian Federation.

25. Ibid., 15.

26. Kyung-Min Kim, "South Korean Capabilities for Space Security," 71.

27. Ibid.

28. Ibid., 67. Also, author's interview with Dr. Chin-Young Hwang, director of the Policy and International Relations Division, KARI, Taejon, South Korea (September 24, 2008).

29. Pinkston, "North and South Korean Space Development," 215–216.

30. Young-Gul Kim, "Innovation and the Role of Korea's Universities," 129. As Kim observed at the time, "Owing to this misguided policy, [South] Korea has a glut of poorly educated scientists and engineers at all levels."

31. Ibid., 136.

32. Victor Cha, "The Security Domain of South Korea's Globalization," in Samuel S. Kim, ed., *Korea's Globalization* (New York: Cambridge University Press, 2000), 224.

33. Ibid., 233.

34. David Wright, "Taepodong 1 test flight" (September 2, 1998), Web site of the Federation of American Scientists, http://www.fas.org/news/dprk/1998/980831-dprk-dcw .htm. Subsequent hypotheses about the Taepodong's potential range—had the third stage actually worked—pushed the range estimates above two thousand kilometers.

35. Lewis Franklin and Nick Hansen, "North Korea's Space Programme," *Jane's Intelligence Review* (September 2009): 13.

36. Daniel A. Pinkston, "The North Korean Ballistic Missile Program," U.S. Army War College report, Carlisle, Penn. (February 2008), http://www.strategicstudiesinstitute .army.mil/pdffiles/PUB842.pdf, p. 19.

37. Tae-Hyung Kim, "South Korea's Space Policy and Its National Security Implications," paper presented at the annual convention of the International Studies Association, New Orleans, La. (February 18, 2010), 18.

38. Kyung-Min Kim, "South Korean Capabilities for Space Security," 69.

39. Ibid.

40. Information on STSat-1 mentioned in "The STSAT-2 (Satellite Technology Satellite 2)," on the KARI Web site, http://new.kari.re.kr/english/02_cms/cms_view .asp?iMenu_seq=119.

41. Korea Aerospace Research Institute, "KARI," organizational booklet (KARI-PUBL-2007–007), 10.

42. Ibid., 70.

43. On this change in the MTCR's rules, see Bowen, *The Politics of Ballistic Missile Proliferation*, 174.

44. Kee, "A South Korean Perspective on Strengthening Space Security in East Asia," 16.

45. Pinkston, "North and South Korean Space Development," 214.

46. Press Release, "Osychshestvlen perviy pusk rakety-nositelya KSLV-1" (First Launch of the KSLV-1 Accomplished), August 28, 2009, Khrunichev Construction

Bureau Web site, http://www.khrunichev.ru/main.php?id=1&nid=1265; Konstantin Lantratov, "Affordable Space Projects from Russia," *Kommersant* (August 21, 2007), http://www.kommersant.com/t795701/r_3/n_25/Space_Rocket_Launch/.

47. "Korea, Russia Enter Full-Fledged Space Partnership," *AsiaPulse News* (July 3, 2007).

48. Ibid.

49. Ministry of Science and Technology, "Reports on Space Activities in the Republic of Korea in 2007," February 2008, submitted to the Committee on the Peaceful Uses of Outer Space, Vienna (February 11–22, 2008) session, p. 13.

50. Cho Jin-seo, "Astronaut Ko San Confirmed for Space Trip," *Korea Times* (October 15, 2007), http://www.koreatimes.co.kr/www/news/tech/2010/01/129_11915.html.

51. "Korea's First Astronaut to Be Replaced," *Korea Times* (March 10, 2008), http://www.koreatimes.co.kr/www/news/tech/2009/10/129_20416.html.

52. "First S. Korean Astronaut Launches," BBC News (April 8, 2008), http://news.bbc.co.uk/2/hi/7335874.stm.

53. Poll figures cited during author's interview with Dr. Chin-Young Hwang, director of the Policy and International Relations Division, KARI, Taejon, South Korea (September 24, 2008).

54. "S. Korean Satellite Lost After Flawed Launch," *Korea Times* (August 26, 2008), http://www.koreatimes.co.kr/www/news/nation/2009/08/113_50711.html.

55. Kim Tong-hyung, "Russia Denies Engine Fault in Rocket Explosion," *Korea Times* (June 13, 2010), http://www.koreatimes.co.kr/www/news/biz/2010/06/123_67579.html.

56. Somi Seong and Steven W. Popper, *Strategic Choices in Science and Technology: Korea in the Era of a Rising China* (Santa Monica, Calif.: RAND, 2005), 103.

57. Ibid., 103; see also chart on 104–105.

58. Ibid., xxvii and xxv.

59. Lee, quoted in "Korea Aims to Bridge Space Tech Gap in 10 Years: KARI Chief," Telecoms Korea Web site, http://www.telecomskorea.com/technology-7433.html.

60. Press Release, "Expanded Satellite Test and Integration Center for Stationary and Large Satellites Construction Complete," May 20, 2008, KARI Web site, http://new.kari.re.kr/english/02_cms/cms_view.asp?iMenu_seq=83&strQueryType=&strQueryWord=&strQueryCategory=&strViewState=view&iSeq=534&iPage=3.

61. Jeong Hae-yoon, "Satrec Initiative's Double Debut," *Korea IT Times* (March 1, 2009), http://www.koreaittimes.com/print/story/satrec-initiatives-double-debut.

62. Emirates News Agency, "EIAST's DubaiSat-1 Launch Marks Successful Implementation of the Strategic Plan of Dubai and the Federal Government" (May 8, 2009), http://www.uaeinteract.com/docs/EIASTs_DubaiSat-1_launch_marks_successful_implementation_of_the_Strategic_Plan_of_Dubai_and_the_Federal_Government/37106.htm.

63. Ibid.; see also Satrec Initiative, "Space Missions: DubaiSat-2," http://www.satreci.com/eng/ds1_1.html?tno=5.

64. Speech by KARI Vice President Dr. Jeong-Joo Rhiu, NASA Ames Research Center, Mountain View, California (July 1, 2009).

65. Korea Aerospace Research Institute, "KARI," p. 11.

66. Kee, "A South Korean Perspective on Strengthening Space Security in East Asia," 16; "Korea's 1st Weather Satellite Launched," *Chosun Ilbo* (June 28, 2010).

67. Korean Broadcasting Service (KBS), "Successful Launch of the Cheollian Multipurpose Satellite and Korea's Dream of Becoming an Aerospace Powerhouse," http://world.kbs.co.kr/english/news/news_commentary_detail.htm?No=18694.

68. Ibid.; KARI, "The COMS program," http://new.kari.re.kr/english/02_cms/cms_view.asp?iMenu_seq=116.

69. Author's interview with Dr. Chin-Young Hwang, director of the Policy and International Relations Division, KARI, Taejon, South Korea (September 24, 2008).

70. Yonhap News, "Korea Must Use 'IT Niche' to Bolster Space Program: IAF Head" (October 12, 2009).

71. Ibid.

72. Jung Sung-ki, "US Willing to Discuss Revising Guidelines on Seoul's Missile Range," *Korea Times* (July 7, 2009), http://www.koreatimes.co.kr/www/news/nation/2009/07/205_48093.html.

73. Ibid.

74. "S. Korea's Aegis Operational for Ship-based Defense," *Space News* (January 5, 2009): 16.

75. However, a South Korean journalist reported in mid-2008 that the ROK Air Force had plans for a potential ground-based ASAT system by 2030, depending on global conditions and government requirements by that time, but there have been no official announcements about such a system. Budget conditions would likely also affect future system development, testing, or deployment decisions. Author's interview with ROK military officer (name withheld), Seoul, South Korea (September 22, 2008).

76. "South Korea Develops Laser Weapons," MISSILETHREAT.com (Claremont Institute), November 10, 2007 (citing Agence France Press and *Chosun Ilbo* sources), http://www.missilethreat.com/archives/id.46/subject_detail.asp.

77. "Glitch Cripples S. Korea's First Military Satellite," *Korea Times* (September 5, 2008). The problem was eventually fixed.

78. "Koreasat 5, a Civil/Military Solutions [*sic*]," Thales Web site, http://www.thalesgroup.com/Case_Studies/CaseStudy_space_Koreasat5/.

79. Press Release, "Orbital Partners with Thales Alenia Space to Build KT Corporation's Koreasat 6 Communications Satellite" (May 14, 2008), on the Orbital Sciences Web site, http://www.orbital.com/NewsInfo/release.asp?prid=656; "Hispasat 1E, Koreasat-6 Launched Aboard Ariane 5," *Space News* (January 10, 2011): 16.

80. Daniel Pinkston, "South Korea to Launch First Military Communication Satellite and Improved Observation Satellite," *WMD Insights* (April 2006), http://www.wmdinsights.com/I4/EA3_SouthKoreatoLaunch.htm.

81. Tae-Hyung Kim, "South Korea's Space Policy and Its National Security Implications," 18.

82. Speech by KARI Vice President Dr. Jeong-Joo Rhiu, NASA Ames Research Center, Mountain View, California (July 1, 2009).

83. "Japan Rocket to Launch South Korean Satellite," *Reuters* (January 12, 2009), http://www.reuters.com/article/2009/01/12/korea-japan-space-idUST11150820090112.

84. Pinkston, "North and South Korean Space Development," 216.

85. Gilbert Rozman, "South Korea and Sino-Japanese Rivalry: A Middle Power's Options Within the East Asian Core Triangle," *Pacific Review* 20, no. 2 (June 2007): 209.

86. For many Koreans, Japan's colonial policies amounted to a strategy of "cultural genocide" against the Korean nation. On this point, see Edward A. Olsen, *Korea: The Divided Nation* (Westport, Conn.: Praeger Security International, 2005), 51.

87. As Heo and Woo argue, changes in the political elite from military leadership to civilian, antigovernment lawyers and activists brought about a "new national identity" with a much less favorable view of the United States, which the new elite identified with decades of hard-line military governments in Seoul. Lee Myung-bak's election changed the elite structure again, bringing in postreform business interests, thanks in part to a perceived failure of the "sunshine policy" and to hostile acts by North Korea from 2006 through 2009. See Uk Heo and Jung-Yeop Woo, "South Korea's Response: Democracy, Identity, and Strategy," in Shale Horowitz, Uk Heo, and Alexander C. Tan, eds., *Identity and Change in East Asian Conflicts: The Cases of China, Taiwan, and the Koreas* (New York: Palgrave, 2007), 157.

88. "Statement by Ambassador Chang, Dong-hee," Plenary Meeting of the Conference on Disarmament, Geneva (August 26, 2008), Web site of the United Nations Office at Geneva, http://www.unog.ch/80256EDD006B8954/(httpAssets)/7E9745CF23F7A445C12574B100383E78/$file/1115_Korea_E.pdf.

89. KARI survey data, cited during author's interview with Dr. Chin-Young Hwang, director of the Policy and International Relations Division, KARI, Taejon, South Korea (September 24, 2008).

90. Kyung-Min Kim, "South Korean Capabilities for Space Security," 69.

91. Author's interview with Dr. Chin-Young Hwang, director of the Policy and International Relations Division, KARI, Taejon, South Korea (September 24, 2008).

92. GlobalSecurity.org, "KITSAT—Korea Institute of Technology Satellite," http://www.globalsecurity.org/space/world/rok/kitsat.htm.

93. Ibid.

94. Speech by KARI Vice President Dr. Jeong-Joo Rhiu, NASA Ames Research Center, Mountain View, California (July 1, 2009).

95. Author's interview with a NASA official in the Office of External Relations, NASA Headquarters, Washington, D.C. (May 8, 2009).

96. Pinkston, "North and South Korean Space Development," 221.

97. Ibid.

98. Huntley, "Smaller State Perspectives on the Future of Space Governance," 254.

99. Press Release (No. 02–2008), "KISS (KOMPSAT-3A Infrared Sensor System)," KARI and AIM INFRAROT-MODULE GmbH (Germany) (February 28, 2008), http://flexfiles.stimme.net/aim-ir.com/admin/_upl_/Aktuelles/Produktneuheiten/Pressemitteilung02_2008_KISS.pdf.

100. Won-hwa Park, "Report: Recent Space Developments in South Korea," *Space Policy* 26, no. 2 (May 2010): 118.

101. Ibid.

102. Speech by KARI Vice President Dr. Jeong-Joo Rhiu, NASA Ames Research Center, Mountain View, California (July 1, 2009).

6. EMERGING ASIAN SPACE PROGRAMS

1. Wade L. Huntley, "Smaller State Perspectives on the Future of Space Governance," *Astropolitics* 5, no. 3 (September–December 2007): 252.

2. Jo-Anne Gilbert, "'We Can Lick Gravity, but . . .': What Trajectory for Space in Australia?" *Space Policy* 25, no. 3 (August 2009): 175.

3. On these developments, see Alice Gorman, "La Terre e l'Espace: Rockets, Prisons, Protests and Heritage in Australia and French Guiana," *Archaeologies: Journal of the World Archaeological Congress* 3, no. 2 (August 2007); Jeff Kingwell, "International Space Year," from the *Year Book Australia, 1992*, posted on the Web site of the Australian Bureau of Statistics, http://www.abs.gov.au/ausstats/abs@.nsf/featurearticlesbytitle/11DE50ACFDEA9948CA2569DE00290F50?OpenDocument#.

4. Gorman, "La Terre e l'Espace," 156, 161.

5. Brett Biddington, personal communication with the author (December 15, 2011).

6. On the growth of these facilities, see Jeffrey T. Richelson, *America's Space Sentinels: DSP Satellites and National Security* (Lawrence: University of Kansas Press, 1999), 53–56.

7. Daphne Burleson, *Space Programs Outside the United States* (Jefferson, N.C.: McFarland and Company, 2005), 10; Jeff Kingwell, "Punching Below Its Weight: Still the Future of Space in Australia?" *Space Policy* 21, no. 2 (May 2005): 162; Australian Space Research Institute (ASRI), "Weapons Research Establishment Satellite (WRESAT)," http://www.asri.org.au/satellites/wresat.

8. Burleson, *Space Programs Outside the United States*, 10.

9. Amsat (Radio Amateur Satellite Corporation), "Australis [*sic*]-Oscar 5," http://www.amsat.org/amsat-new/satellites/satInfo.php?satID=28&retURL=satellites/frequencies.php.

10. Kingwell, "Punching Below Its Weight," 162.

11. Australian Communications and Media Authority (ACMA), "Australian Satellite Networks," http://www.acma.gov.au/WEB/STANDARD/pc=PC_533.

12. The author thanks Jim Vedda for this information.

13. "Australia Signs Space Launch Agreement with Russia" *Space Daily* (May 23, 2001), http://www.spacedaily.com/news/aust-01a.html.

14. Kingwell, "Punching Below Its Weight," 162.

15. NASA, "NASA to Open New Competition for Space Transportation Seed Money," press release (October 18, 2007); see also Burleson, *Space Programs Outside the United States*, 16.

16. "Australian Government Space Engagement: Policy Framework and Overview" (2003), cited in Kingwell, "Punching Below Its Weight," 161.

17. Kingwell, "Punching Below Its Weight," 162.

18. Australian Senate, Standing Committee on Economics, "Lost in Space? Setting a New Direction for Australia's Space Science and Industry Sector," Senate Printing Unit, Canberra (November 2008), 1.

19. Australian Government, "Government Response to the Inquiry By the Senate Standing Committee on Economics Into the Current State of Australia's Space Science and Industry Sector" (November 2009).

20. Gilbert, "'We Can Lick Gravity, but . . . ,'" 175. The figure in Australian dollars was $160.5 million.

21. See Australian Space Research Program, "Program Guidelines—October 2009," on the Web site of the Australian government's Department of Innovation, Industry, Science and Research, http://www.space.gov.au/AustralianSpaceResearchProgram/Documents/ASRP_Guidelines_October_2009.pdf.

22. Brett Biddington, "Remote-Sensing Essential for Security" (commentary), *Weekend Australian* (October 24, 2009): 12.

23. Aaron Fernandes, "Zadko to Expand Australia's Space Program in 2010" (January 28, 2010), Western Australia Science Web site, http://www.sciencewa.net.au/index.php?option=com_content&view=article&id=2942:zadko-to-expand-australias-space-program-in-2010&catid=201:news&Itemid=200079.

24. Brett Biddington, personal communication with the author (December 15, 2011).

25. Biddington, "An Australian Perspective on Space Security," 102.

26. Peter B. de Selding, "Australian Satellite Broadband Plan Reaches Into Suburbia," *Space News* (July 19, 2010): 5.

27. Peter B. de Selding, "Australian, U.S. Forces to Share UHF Satellite Capacity," *Space News* (May 3, 2010): 7.

28. Biddington, "An Australian Perspective on Space Security," 101; Biddington, "Remote-Sensing Essential for Security."

29. Boeing corporation, "Boeing Awarded $234 Million to Complete Production of 6th Wideband Global SATCOM Satellite," press release (January 15, 2009), http://www.boeing.com/news/releases/2009/q1/090115b_nr.html; "US Air Force Readies for Third WGS Satellite to Extend Warfighter Capabilities," *Satnews Daily* (November 30, 2009), http://www.satnews.com/cgi-bin/story.cgi?number=289583046.

30. Biddington, "An Australian Perspective on Space Security," 99.

31. Brett Biddington and Roy Sach, *Australia's Place in Space: Toward a National Space Policy*, Kokoda Foundation (Australia), Kokoda Paper no. 13 (June 2010).

32. K. K. Nair, *Space: The Frontiers of Modern Defence* (New Delhi: Knowledge World, 2006), 190.

33. Ibid., 187.

34. Ibid., 188.

35. Ibid.

36. Burleson, *Space Programs Outside the United States*, 148.

37. Steven Lambakis, *On the Edge of Earth: The Future of American Space Power* (Lexington: University Press of Kentucky, 2001), 156.

38. Burleson, *Space Programs Outside the United States*, 147.

39. Ibid., 148.

40. Nair, *Space: The Frontiers of Modern Defence*, 189; Institut fur Luft-und Raumfahrt, Technische Universitat Berlin, Raumfahrttechnik, "Overview," http://www.raumfahrttechnik.tu-berlin.de/tubsat/lapan-tubsat/.

41. "Reshetnev Concludes Deal for Indonesia's Telecom-3," *Space News* (March 16, 2009): 5.

42. Peter B. de Selding, "Palapa-D to Be Salvaged After Being Launched Into Wrong Orbit," *Space News* (September 7, 2009): 10.

43. Astronautix.com, "RX-250" and "RX-250 Chronology," http://www.astronautix.com/lvs/rx250.htm.

44. APRSAF, "LAPAN Launched RX-420 Rocket," *APRSAF Newsletter* 57 (August 28, 2009), http://www.aprsaf.org/newsmails_newsletters/mails_2009/newsmail057.php.

45. "Space Agency Launches RX-420 Rocket," *Jakarta Globe* (July 2, 2009).

46. Burleson, *Space Programs Outside the United States*, 193.

47. International Astronautical Federation (IAF), "Interview with Dr Mazlan Othman," http://www.iafastro.net/index.php?id=564.

48. The Malaysian National Space Agency Web site, http://www.angkasa.gov.my/.

49. Burleson, *Space Programs Outside the United States*, 193.

50. Nair, *Space: The Frontiers of Modern Defence*, 191.

51. Burleson, *Space Programs Outside the United States*, 194.

52. Nair, *Space: The Frontiers of Modern Defence*, p. 191; Angkasa, "TiungSAT-1," http://www.angkasa.gov.my/index2.php?option=com_content&do_pdf=1&id=106.

53. Ibid.

54. Ibid.

55. Associated Press, "Malaysia Targets 2020 for Possible Manned Moon Mission" (August 25, 2005); Agence France Presse, "Malaysia's 25 Million Dollar Astronaut Program on Track: Official" (August 23, 2005).

56. Prime Minister Mahathir Mohamed, quoted in IAF, "Interview with Dr Mazlan Othman."

57. Hazlin Hassan, "The First Malaysian in Space," *Cosmos* (June 8, 2006), http://www.cosmosmagazine.com/features/online/337/the-first-malaysian-space.

58. Tariq Malik, "SpaceX Successfully Launches Commercial Satellite to Orbit," *Space News* (July 20, 2009): 20; Stephen Clark, "Commercial Launch of SpaceX Falcon 1 Rocket a Success," *Spaceflight Now* (July 14, 2009), http://spaceflightnow.com/falcon/005/index.html.

59. Malaysian National News Agency, "RazakSat Successfully Blasts Off into Space," press release (July 14, 2009).

60. On these goals, see "Statement by the Hon. Ahmad Shabery Cheek," Malaysian member of Parliament, delivered at the Fourth Committee of the Fifty-ninth Session of the UN General Assembly, New York (October 18, 2004).

61. Lewis Franklin and Nick Hansen, "North Korea's Space Programme," *Jane's Intelligence Review* (September 2009): 12.

62. Ibid., 13.

63. BBC News, "Defiant N Korea launches rocket" (April 5, 2009).

64. Daniel A. Pinkston, "North and South Korean Space Development: Prospects for Cooperation and Conflict," *Astropolitics* 4, no. 2 (Summer 2006): 219.

65. Terence Roehrig, "North Korea in Crisis: Regime, Identity, and Strategy," in Shale Horowitz, Uk Heo, and Alexander C. Tan, eds., *Identity and Change in East Asian Conflicts: The Cases of China, Taiwan, and the Koreas* (New York: Palgrave, 2007), 142.

66. Robert H. Schmucker and Markus Schiller, "The DPRK Missile Show: A Comedy in (Currently) Eight Acts," draft paper by Schmucker Technologie (Munich, Germany; May 5, 2010).

67. On this cooperation, see Global Security Newswire, "Iran, North Korea Seen Collaborating on Rocket Launchpad" (March 8, 2010), available on the Nuclear Threat Initiative Web site, http://www.globalsecuritynewswire.org/gsn/nw_20100308_9158.php.

68. Author's interviews with faculty from the Beijing University of Aeronautics and Astronautics, Beijing, China (September 2009).

69. Mike Chinoy, *Meltdown: The Inside Story of the North Korean Nuclear Crisis* (New York: St. Martin's Griffin, 2009), 378.

70. On A. Q. Khan's activities, see David Albright, *Peddling Peril: How the Secret Nuclear Trade Arms America's Enemies* (New York: Free Press, 2010).

71. Nair, *Space: The Frontiers of Modern Defence*, 133.

72. SUPARCO, "History," http://www.suparco.gov.pk/pages/history.asp.

73. Nair, *Space: The Frontiers of Modern Defence*, 133.

74. Ibid., 135.

75. Ibid., 136.

76. Burleson, *Space Programs Outside the United States*, 213.

77. Ibid.

78. Nair, *Space: The Frontiers of Modern Defence*, 138.

79. Ibid., 139.

80. Globalsecurity.org, "Small Multimission Spacecraft (SMMS)," http://www.globalsecurity.org/space/world/china/smms.htm.

81. Nair, *Space: The Frontiers of Modern Defence*, 140.

82. SUPARCO, "Remote Sensing Satellite," http://www.suparco.gov.pk/pages/rsss.asp.

83. Nair, *Space: The Frontiers of Modern Defence*, 141.

84. SUPARCO, "Paksat-1R," http://www.suparco.gov.pk/pages/paksat1r.asp?satlink sid=1; Great Wall Industry Corporation (GWIC), "Paksat-1R Program," http://www.cgwic.com/In-OrbitDelivery/CommunicationsSatellite/Program/PakSat-1R.html.

85. "Chinese-built Paksat-1R to Include Pakistani Payloads," *Space News* (December 14, 2009): 12.

86. Indonesia Arab Blog, "Pakistan Plans to Launch Its Own Satellite" (September 17, 2008; posted by "Ngeditor, Pakistan"), http://indonesiaarab.wordpress.com/2008/09/17/pakistan-plans-to-launch-its-own-satellite/.

87. Nair, *Space: The Frontiers of Modern Defence*, 136.

88. Ibid., 137.

89. Indonesia Arab Blog, "Pakistan Plans to Launch Its Own Satellite."

90. SUPARCO, "International Cooperation: Inter-Islamic Network on Space Sciences and Technology (ISNET)," http://www.suparco.gov.pk/pages/isnet.asp.

91. Philippine Atmospheric, Geophysical, and Astronomical Administration (PAGASA), "The 'Observatorio,'" http://www.pagasa.dost.gov.ph/fullhistory.shtml.

92. Ibid.

93. Isidro S. Fajardo, "Surveying and Mapping in the Philippines," paper presented at the Twentieth International Cartographic Conference, Beijing, China (August 6–10, 2001).

94. Ibid.

95. Burleson, *Space Programs Outside the United States*, 215.

96. Helen Flores, "Move Over, NASA, Here Comes PASA," *Philippine Star* (January 8, 2007).

97. Resource Recovery Movement, posting entitled "Philippine Space Agency," October 21, 2009, http://ecologyguardian.wordpress.com/.

98. See the MSC Web site, http://www.mabuhaysat.com.

99. Emmie V. Abadilla, "Asia Broadcast Satellite Establishes RP as Regional Satellite Hub," Manila Bulletin Publishing Corporation (MBPC) (February 28, 2010), http://www.mb.com.ph/articles/245488/asia-broadcast-satellite-establishes-rp-regional-satellite-hub. The *Agila-1* satellite failed to achieve orbit.

100. Space Systems/Loral, "Agila 2," http://www.ssloral.com/downloads/products/agila2.pdf.

101. Asia Broadcast Satellite (ABS), "Asia Broadcast Satellite Acquires Koreasat-3" (May 24, 2010), http://www.absatellite.net/news/news_100424.html.

102. Jermyn Chow, "First Made-in-S'pore Satellite to Launch Soon," *Straits Times* (March 29, 2010): A6.

103. Ibid.

104. Indian Space Research Organisation, "PSLV to Launch Singapore University Satellite," press release (February 4, 2003).

105. Satnews.com, "Singapore, Xpect X-Sat" (March 28, 2010), http://www.satnews.com/cgi-bin/story.cgi?number=2019349931.

106. Timo Bretschneider, "Singapore's Satellite Mission X-Sat," Nanyang Technological University, Singapore, paper presented at Fourth IAA Symposium on Small Satellites for Earth Observation, Berlin, Germany (April 7–11, 2003).

107. Ibid.

108. Richard Leshner, "Satellite Communications and the Pacific Rim," in Rebecca Jimerson and Ray A. Williamson, eds., *Space and Military Power in East Asia: The Challenges and Opportunity of Dual-Purpose Space Technologies*, a report of the Space Policy Institute, George Washington University, 2001, available on the Space Policy Institute Web site, http://www.gwu.edu/%7Espi/assets/docs/spacemiltoc.html.

109. Ibid.; David Bruggeman, "Remote Sensing in East Asia," in Rebecca Jimerson and Ray A. Williamson, eds., *Space and Military Power in East Asia: The Challenges and Opportunity of Dual-Purpose Space Technologies*, a report of the Space Policy Institute, George Washington University, 2001, available on the Space Policy Institute Web site, http://www.gwu.edu/%7Espi/assets/docs/spacemiltoc.html.

110. National Space Organization (NSPO), "Formosat-2: Program Description," http://www.nspo.org.tw/2008e/projects/project2/intro.htm.

111. Agence France Presse, "Taiwan 'Borrows' Israeli Satellite During Overpasses" (August 12, 2001).

112. Ibid.

113. NSPO, "Heritage," http://www.nspo.org.tw/2008e/aboutNSPO/origin.htm.

114. NSPO, "Sounding Rocket Program," http://www.nspo.org.tw/2008e/projects/projectrocket/intro.htm.

115. NSPO, "Formosat-3: Program Description," http://www.nspo.org.tw/2008e/projects/project3/intro.htm.

116. NSPO, "ESEMS: Program Description," http://www.nspo.org.tw/2008e/projects/projectESEMS/intro.htm.

117. "SpaceX, Falcon 1e to Launch Taiwan's Formosat-5 Craft," *Space News* (June 21, 2010): 9.

118. SpaceTech GmbH Immenstaad (STI), "ARGO/Formosat-5 Mission Design and Hardware Kit Procurement," http://www.spacetech-i.com/ARGO.html.

119. Geo-Informatics and Space Technology Development Agency (GISTDA), "Profile," http://www.gistda.or.th/gistda_n/en/index.php?option=com_content&view=article&id=5&Itemid=2.

120. APRSAF, "Interview with Dr. Thongchai Charuppat of the Geo-Informatics and Space Technology Development Agency" (January 27, 2010), Bangkok, Thailand, at the APRSAF-16 conference, http://www.aprsaf.org/interviews_features/interviews_2010/40.php.

121. GISTDA, "Profile."

122. Ibid.

123. Burleson, *Space Programs Outside the United States*, 299.

124. Thaicom Public Company Limited, "Thaicom 3," http://www.thaicom.net/eng/satellite_thaicom3.aspx.

125. Space Systems/Loral, "Loral-Built Thaicom 4 (IPStar) Successfully Launched" (August 11, 2005), http://www.ssloral.com/html/pressreleases/pr20050811.html.

126. De Selding, "Australian Satellite Broadband Plan Reaches Into Suburbia."

127. APRSAF, "Interview with Dr. Thongchai Charuppat."

128. Asian Surveying & Mapping (ASM), "Thailand Launches Theos" (October 2, 2008), http://www.asmmag.com/news/thailand-launches-theos (accessed July 29, 2010).

129. APRSAF, "Interview with Dr. Thongchai Charuppat."

130. Dr. Surachai Ratanasermpong (director, Geo-Informatics Office, GISTDA), "Space Technology and Applications in Thailand," paper presented at APRSAF-12

conference, Kitakyushu, Japan (October 11–12, 2005); ASM, "Thailand Launches Theos."

131. APSCO, "History," http://www.apsco.int/history.aspx.

132. Peter B. de Selding, "Thailand's Political Unrest Has Thaicom Caught in the Middle," *Space News* (May 3, 2010): 6.

133. Mondhon Snaguansermsri, "The Space Technology and Geo-Informatics Program at Naresuan University, Thailand," paper presented at the International Workshop on "Processing and Visualization Using High-Resolution Imagery," Pitsanulok, Thailand (November 18–20, 2004).

134. APRSAF, "Interview with Dr. Thongchai Charuppat."

135. Nong Khac Y, "Vietnam's Space Technology: Starting with 1kg Satellite" (interview with Dr. Pham Anh Tuan, associate dean of the Space Technology Institute of Vietnam), *Hanoi Times* (November 22, 2007), http://www.hanoitimes.com.vn/print .asp?newsid=741.

136. Ibid.

137. Mai Luan, "Vietnam About to Launch Supersmall Satellite" (September 17, 2009), Lookatvietnam.com Web site (a Vietnamese news daily), http://www.lookat-vietnam.com/2009/09/vietnam-about-to-launch-super-small-satellite.html.

138. VietNam.Net Bridge, "Vinasat-1 Satellite Goes to Launching Pad" (April 18, 2008), http://english.vietnamnet.vn/tech/2008/04/778819/.

139. Lockheed Martin Corporation, "Lockheed Martin–Built Vinasat-1 Satellite Launched Successfully for Vietnam Telecommunications Group," press release (April 18, 2008).

140. Xinhua News Agency, "Vietnam Establishes Space Technology Institute" (April 7, 2007).

141. "Arianespace Picked to Loft Vietnam's Vinasat-2 Satellite," *Space News* (June 21, 2010): 9.

142. Tran Manh Tuan, "Space Technology in Vietnam: 2008 Country Report," presentation at the APRSAF-15 conference, "Space for Sustainable Development," Hanoi and HaLong Bay, Vietnam (December 9–12, 2008).

143. Ibid.

144. Doan Minh Chung, "Vietnam Country Report: Space Technology Activities in 2009," presentation at the APRSAF-16 conference, "Space Applications: Contributions Towards Human Safety and Security," Bangkok, Thailand (January 26–29, 2010).

145. Ibid.

146. For more information, see the Web site of the Vietnam Space Technology Institute: http://www.sti.vast.ac.vn/ (accessed July 29, 2010).

147. Xinhua News Agency, "Vietnam Establishes Space Technology Institute."

148. Ibid.

149. Doan, "Vietnam Country Report: Space Technology Activities in 2009."

150. Vietnam.Net Bridge/Viet Nam News, "VN Space Institute to Make Small Satellites" (June 25, 2010), http://english.vietnamnet.vn/tech/201006/VN-space-institute -to-make-small-satellites-918313/.

151. Globalsecurity.org, "Vinasat," http://www.globalsecurity.org/space/world/vietnam/comm.htm.

152. Nhan Dan, "Vinasat-1 First of Many Says Vietnam," *Vietnam News Agency* (January 24, 2009).

153. Vietnam Press Center (VPC), "Vinasat-2 Satellite Launches Early" (interview with Hoang Minh Thong, director of the Vietnam Post and Telecommunications Group's telecom project management unit; January 12, 2010), http://www.presscenter.org.vn/en/content/view/1921/89/.

154. Nhan, "Vinasat-1 First of Many Says Vietnam."

155. Nong, "Vietnam's Space Technology: Starting with 1kg Satellite" (interview with Dr. Pham Anh Tuan).

156. Ibid.

157. Doan, "Vietnam Country Report: Space Technology Activities in 2009."

7. ASIA'S SPACE RACE: IMPLICATIONS FOR REGIONAL AND GLOBAL POLICY

1. Mely Caballero-Anthony, "Nontraditional Security and Multilateralism in Asia: Reshaping the Contours of Regional Security Architecture," in Michael J. Green and Bates Gill, eds., *Asia's New Multilateralism: Cooperation, Competition, and the Search for Community* (New York: Columbia University Press, 2010), 322.

2. See Alexander Gerschenkron, *Economic Backwardness in Historical Perspective: A Book of Essays* (Cambridge, Mass.: Harvard University Press, 1962).

3. Muthiah Alagappa, "Introduction: Predictability and Stability Despite Challenges," in Muthiah Alagappa, ed., *Asian Security Order: Instrumental and Normative Features* (Stanford, Calif.: Stanford University Press, 2003), 17.

4. Bates Gill and Michael J. Green, "Unbundling Asia's New Multilateralism," in Michael J. Green and Bates Gill, eds., *Asia's New Multilateralism: Cooperation, Competition, and the Search for Community* (New York: Columbia University Press, 2010), 12.

5. The exceptions were the apparent finding by U.S. instruments on the *Chandrayaan* of evidence of water ice on the Moon and the higher-quality mapping of the Moon by the Japanese *Kaguya* mission, thanks to its use of HD imaging technology.

6. Kiran Nair, "Space Security: Reassessing the Situation and Exploring Options," in John M. Logsdon and James Clay Moltz, eds., *Collective Security in Space: Asian Perspectives* (Washington, D.C.: George Washington University, Space Policy Institute, January 2008), 93.

7. Ibid.

8. Roger D. Launius, "United States Space Cooperation and Competition: Historical Reflections," *Astropolitics* 7, no. 2 (May–August 2009): 98.

9. Ibid.

10. ASPCO, "APSCO Meet with JAXA," undated report, http://www.apsco.int/newsCont.aspx?news_id=11038.

11. Roger Handberg and Zhen Li, *Chinese Space Policy: A Study in Domestic and International Politics* (New York: Routledge, 2007), 170.

12. D. Narayana Moorthi, "What 'Space Security' Means to an Emerging Space Power," *Astropolitics* 2, no. 2 (Summer 2004): 268.

13. Ibid.

14. "Indian Statement on Satellite TV Ownership Falls Well Short of a Breakthrough" (Editorial), *Space News* (July 26, 2010): 18.

15. K. K. Nair, *Space: The Frontiers of Modern Defence* (New Delhi: Knowledge World, 2006), 129.

16. Lt. Col. (USAF) James Mackey, "Recent US and Chinese Antisatellite Activities," *Air and Space Power Journal* 23, no. 3 (Fall 2009): 91.

17. Ibid.

18. Thom Shanker, "China's Military an Obstacle to Improving Relations, Gates Says," *New York Times* (June 4, 2010).

19. "Gates Sees 'Disconnect' Between Chinese Civilian, Military Leaders," *Global Security Newswire* (January 14, 2011), on the Nuclear Threat Initiative Web site, http://www.globalsecuritynewswire.org/gsn/nw_20110114_4025.php.

20. Kevin Pollpeter, "Building for the Future: China's Progress in Space Technology During the Tenth 5-Year Plan and the U.S. Response," U.S. Army War College, Strategic Studies Institute (March 2008), viii.

21. John M. Logsdon, "Why Space Exploration Should Be a Global Project," *Space Policy* 24, no. 1 (February 2008): 3.

22. Ibid., 5.

23. Moorthi, "What 'Space Security' Means to an Emerging Space Power," 262.

24. Ibid., 266.

25. Roger Handberg and Zhen Li, *Chinese Space Policy: A Study in Domestic and International Politics* (New York: Routledge, 2007), 170.

26. Dean Cheng, "U.S.-China Space Cooperation: More Costs Than Benefits," WebMemo, Heritage Foundation, No. 2670 (October 30, 2009), 1.

27. Eric Sterner, "U.S.-China Space Relations: Maintaining an Arm's Length," *Space News* (March 2, 2009): 19.

28. Culberson's letter, as excerpted in *Space News* (October 18, 2010): 18.

29. Bruce Bade and William O. Berry, "Engaging China in Science and Technology: A Wise Endeavor for the Department of Defense," *Common Defense Quarterly* (Fall 2009).

30. Gregory Kulacki and Jeffrey G. Lewis, "Report: Understanding China's Antisatellite Test," *Nonproliferation Review* 15, no. 2 (July 2008): 344.

31. Ibid.

32. Peter B. de Selding, "APT Bucks Consolidation Trends in Asia," *Space News* (July 5, 2010): 13.

33. Ibid.

34. Susan L. Shirk, *China: Fragile Superpower* (New York: Oxford University Press, 2008), 268.

35. Alanna Krolikowski, "China's Civil and Commerical Space Activities and Their Implications," testimony before the U.S.–China Economic and Security Review Commission, Washington, D.C., May 11, 2011, 12.

36. Ken Pole, "U.S. Defense Companies Face ITAR Challenges," *Frontline* (Ottawa, Canada) 7, no. 2 (March/April 2010): 9–10.

37. Phillip C. Saunders, "Managing Strategic Competition with China," Strategic Forum (National Defense University's Institute for National Strategic Studies) no. 242 (July 2009): 6.

38. Jay Solomon, "U.S. Takes on Maritime Spats: Clinton Plan Would Set up Legal Process for Asian Nations to Resolve Claims in the South China Sea," *Wall Street Journal* (July 24–25, 2010): A8.

39. White House, "National Space Policy of the United States of America" (June 28, 2010), on the White House Web site, http://www.whitehouse.gov/sites/default/files/national_space_policy_6-28-10.pdf.

40. Logsdon, cited in Amy Klamper, "International Cooperation Emphasis of Obama Space Policy," *Space News* (July 5, 2010): 6.

41. President Obama, quoted in Ibid.

42. Nina-Louisa Remuss, "The Fair and Responsible Use of Space: An International Perspective," *Space Policy* 25, no. 1 (February 2009): 64.

43. Per Magnus Wijkman, "Managing the Global Commons," *International Organization* 36, no. 3 (Summer 1982): 535.

44. Doo Hwan Kim, "Korea's Space Development Programme: Policy and Law," *Space Policy* 22, no. 1 (February 2006).

45. Ibid., 117.

46. Remuss, "The Fair and Responsible Use of Space," 64.

47. Moorthi, "What 'Space Security' Means to an Emerging Space Power," 262.

48. James A. Vedda, *Choice, Not Fate: Shaping a Sustainable Future in the Space Age* (XLibris, 2009).

49. Ibid., 181.

50. Ibid., 186.

51. Pascale Ehrenfreund and Nicolas Peter, "Managing Future International Space Exploration Activities: Toward a Paradigm Shift," *Space News* (April 27, 2009): 21.

52. Ibid., 22.

53. Michael D. Griffin, NASA Chief Administrator, memorandum on "Chinese Lunar Capabilities" (November 13, 2007), 4.

54. Andy Pasztor, "China Sets Ambitious Space Goals," *Wall Street Journal* (April 15, 2010): A4.

55. Author's interview with Chinese space analyst, Beijing, China (September 2009).

56. Yi Zhou, "Perspectives on Sino-US Cooperation in Civil Space Programs," *Space Policy* 24, no. 3 (August 2008): 137.

57. Space Foundation, *The Space Report: 2011* (Colorado Springs: Space Foundation, 2011), 56.

58. "White House Calls for Export Control Agency," *Space News* (July 5, 2010): 3.

59. James A. Lewis, "China as a Military Space Competitor," in John M. Logsdon and Audrey M. Schaffer, eds., *Perspectives on Space Security* (Washington, D.C.: Space Policy Institute, George Washington University, December 2005), 114.

60. For an example of this argument regarding the inevitability of space conflict, see Everett C. Dolman, *Astropolitik: Classical Geopolitics in the Space Age* (London: Frank Cass, 2002).

61. Dennis J. Blair, "Military Power Projection in Asia," in Ashley J. Tellis, Mercy Kuo, and Andrew Marble, eds., *Strategic Asia 2008–09: Challenges and Choices* (Seattle, Wash.: National Bureau of Asian Research, 2008), 419.

62. James Mulvenon, "China: Conditional Compliance," in Muthiah Alagappa, ed., *Coercion and Governance: The Declining Political Role of the Military in Asia* (Stanford, Calif.: Stanford University Press, 2001), 334.

63. Ibid.

64. On this initiative, see Peter B. de Selding, "Satellite Operators Solicit Bids to Create Orbital Database," *Space News* (November 23, 2009): 6.

65. James Michael Snead, "Spacefaring Logistics Infrastructure: The Foundation of a Spacefaring America," *Astropolitics* 6, no. 1 (January–April 2008): 72.

66. Ibid., 79.

67. Quoted in Associated Press, "China Warns of 'Arms Race' in Outer Space" (August 12, 2009).

68. Ibid.

69. Joan Johnson-Freese, *Heavenly Ambitions: America's Quest to Dominate Space* (Philadelphia: University of Pennsylvania Press, 2009), 65.

70. Deborah Welch Larson and Alexei Shevchenko, "Status Seekers: Chinese and Russian Responses to U.S. Primacy," *International Security* 34, no. 4 (Spring 2010).

71. Eligar Sadeh, "Report: United States–China Space Dialogue Project," *Astropolitics* 8, no. 1 (January–April 2010): 12.

72. Ibid., 16.

73. Ibid., 12.

74. The White House, "National Space Policy of the United States of America," 9.

75. Kazuto Suzuki, "Japanese Steps Toward Regional and Global Confidence Building," in John M. Logsdon and James Clay Moltz, eds., *Collective Security in Space: Asian Perspectives* (Washington, D.C.: George Washington University, Space Policy Institute, January 2008).

76. Ibid., 144.

77. Caballero-Anthony, "Nontraditional Security and Multilateralism in Asia: Reshaping the Contours of Regional Security Architecture," 322.

78. Detlev Wolter, *Common Security in Outer Space and International Law* (Geneva: UNIDIR, 2006), 205.

79. On this proposal, see ibid., 150–164.

80. Emily O. Goldman and Andrew L. Ross, "The Diffusion of Military Technology and Ideas—Theory and Practice," in Emily O. Goldman and Leslie C. Eliason, eds., *The*

Diffusion of Military Technology and Ideas (Stanford, Calif.: Stanford University Press, 2003), 401.

81. Ajey Lele, "Space Security and India," in Sukhvinder Kaur Multani, ed., *Space Security* (Hyderabad: Icfai University Press, 2008), 181.

82. Remuss, "The Fair and Responsible Use of Space," 64.

83. Henry A. Kissinger, "Avoiding a U.S.-China Cold War" (op-ed), *Washington Post* (January 14, 2011).

84. Ibid.

85. Moorthi, "What 'Space Security' Means to an Emerging Space Power," 262.

86. Robert Jervis, *Perception and Misperception in International Politics* (Princeton, N.J.: Princeton University Press, 1976), 220.

87. Robert O. Keohane, *After Hegemony: Cooperation and Discord in the World Political Economy* (Princeton, N.J.: Princeton University Press, 1984), 259.

88. Caballero-Anthony, "Nontraditional Security and Multilateralism in Asia," 324.

INDEX

CPSIA information can be obtained
at www.ICGtesting.com
Printed in the USA
LVOW12s0836270318
571307LV00001B/4/P